Atlantis Briefs in Differential Equations

Volume 3

Series editors

Zuzana Došlá, Brno, Czech Republic
Šárka Nečasová, Prague 1, Czech Republic
Milan Pokorný, Praha 8, Czech Republic

About this Series

The aim of the series is rapid dissemination of new results and original methods in the theory of Differential Equations, including topics not yet covered by standard monographs. The series features compact volumes of 75–200 pages, written in a concise, clear way and going directly to the point; the introductory material should be restricted to a minimum or covered by suitable references.

For more information on this series and our other book series, please visit our website at: www.atlantis-press.com/publications/books

AMSTERDAM—PARIS—BEIJING
ATLANTIS PRESS
Atlantis Press
29, avenue Laumière
75019 Paris, France

More information about this series at http://www.springer.com/series/13609

Šárka Nečasová · Stanislav Kračmar

Navier–Stokes Flow Around a Rotating Obstacle

Mathematical Analysis of its Asymptotic Behavior

ATLANTIS
PRESS

Šárka Nečasová
Department of Evolutionary Equations
Mathematical Institute
Academy of Sciences
Prague 1
Czech Republic

Stanislav Kračmar
Department of Technical Mathematics
Czech Technical University
Prague 2
Czech Republic

ISSN 2405-6405   ISSN 2405-6413 (electronic)
Atlantis Briefs in Differential Equations
ISBN 978-94-6239-230-4   ISBN 978-94-6239-231-1 (eBook)
DOI 10.2991/978-94-6239-231-1

Library of Congress Control Number: 2016951702

Printed on acid-free paper

To the memory of Prof. Jindřich Nečas

Preface

The book is devoted to the mathematical analysis of the asymptotic behavior of the motion of viscous fluid around rotating and translating bodies. The work is based on the articles published during 2010–2016 which we were doing together with Paul Deuring. We would like to thank him for his wonderful collaboration and support for this project. We only regret that he could not join our project.

Š.N. would like to express her deep thanks to Prof. G.P. Galdi for introducing her to this wonderful subject. Second, she would like to thank her family—her mother Zdeňka, her sister Jindra and children—Martin, Jan and Lucie for their great support. S.K. would like to thank his wife Dagmar for her support and patience.

Š.N. and S.K. want to express their gratitude to Prof. K. Segeth, who have read the manuscript and contributed to its improvement.

The work of Š. Nečasová and S. Kračmar was supported by Grant No. 16-03230S of the Czech Science Foundation in the framework of RVO 67985840.

Prague 1, Czech Republic   Šárka Nečasová
Prague 2, Czech Republic   Stanislav Kračmar

Contents

Chapter 1
Introduction

Many interesting phenomena deal with a fluid interacting with a moving rigid or deformable structure. These types of problems have a lot of important applications in biomechanics, hydroelasticity, sedimentation, etc.

From the mathematical point of view the problem has been studied over last 40 years. We will focus on the study of Navier–Stokes fluid flows past a rigid body translating with a constant velocity and a rotating with a prescribed constant angular velocity. A systematic and rigorous mathematical study was initiated by the fundamental pioneer works of Oseen (1927), Leray (1933, 1934) and then developed by several other mathematicians with significant contributions in the case of zero angular velocity.

In the last decade a lot of efforts have been made in the analysis of solutions to different problems, stationary as well non-stationary, linear models as well nonlinear ones, in the whole space as well in exterior domains, in the case with prescribed constant angular velocity or angular velocity dependent on time.

We will study the problem of a rigid body $\mathfrak{D}$ translating with constant velocity and rotating with constant angular velocity in an incompressible viscous fluid, i.e. the flow field $\mathcal{F}$ around this body. We consider a rigid body $\mathfrak{D}$ as an open, bounded set with smooth boundary.

Let $V = V(y, t)$ be the velocity field associated with the motion of the body $\mathfrak{D}$ with respect to an inertial frame $\mathcal{I}$ with origin $\mathcal{O}$. Denoting by $y_C = y_C(t)$ the path of the center of mass of $\mathfrak{D}$ and by $\tilde{\omega} = \tilde{\omega}(t) \in \mathbb{R}^3$ the angular velocity of $\mathfrak{D}$ around its center of mass, we have

$$V(y, t) = \dot{y}_C(t) + \tilde{\omega}(t) \times (y - y_C(t)), \tag{1.1}$$

where $\dot{y}_C = dy_C/dt$ is the translational velocity of $\mathfrak{D}$ and, for simplicity, $y_C(0) = 0$.

Let the Eulerian velocity field and pressure associated with the motion of the liquid in $\mathcal{I}$ be denoted by $v = v(y, t)$ and $q = q(y, t)$, respectively. The equations

© Atlantis Press and the author(s) 2016
Š. Nečasová and S. Kračmar, *Navier–Stokes Flow Around a Rotating Obstacle*,
Atlantis Briefs in Differential Equations 3, DOI 10.2991/978-94-6239-231-1_1

of conservation of linear momentum and mass of the fluid are then modeled by the
Navier–Stokes equations.[1]

Given a kinematic viscosity $\nu > 0$ and an external force $\tilde{f} = \tilde{f}(y, t)$, the
unknowns v, q solve the nonlinear system

$$
\begin{aligned}
\partial_t v - \nu \Delta v + (v \cdot \nabla) v + \nabla q &= \tilde{f} & &\text{in } \Sigma(t),\ t \in (0, \infty), \\
\operatorname{div} v &= 0 & &\text{in } \Sigma(t),\ t \in (0, \infty), \\
v(y, t) &= V(y, t) & &\text{on } \partial \Sigma(t),\ t \in (0, \infty), \\
v(y, t) &\to 0 & &\text{as } |y| \to \infty,
\end{aligned} \tag{1.2}
$$

in a time-dependent exterior domain $\Sigma(t) \subset \mathbb{R}^3$.

However, this formulation has an undesired behavior, the region occupied by
$\mathcal{F}$ is an unknown function of time. To deal with this type of problem, you can
introduce a new coordinate system attached to the body. There are two possibilities
of transformation:

- *Global transformation*, which was introduced by Weinberger [57, 58] in the case
  of body falling by gravity in viscous incompressible fluid,or
- *Local transformation*, see Hieber et al. [13].

Let us mention that this type of problem is a particular case of fluid-structure inter-
action, where the motion of body satisfies the Newton law and the conservation of
angular momentum.

We will discuss the case of a time-independent angular velocity $\tilde{\omega} = k e_3$ and
constant translational velocity $0 \neq \dot{y}_C = u_\infty \in \mathbb{R}^3$ so that $y_C(t) = u_\infty t$. For this
reason we introduce the change of variables (global transformation), we denote the
new frame by $\mathcal{J}$, where the origin of $\mathcal{I}$ and $\mathcal{J}$ coincide with the center of mass C of
the body $\mathfrak{D}$ and we assume that $\mathcal{I} = \mathcal{J}$ at time $t = 0$. Thus introducing

$$
x = \mathcal{Q}(t)^T \left(y - y_C(t) \right) \tag{1.3}
$$

and the new functions

$$
u(x, t) = \mathcal{Q}(t)^T v(y, t), \quad p(x, t) = q(y, t), \quad f(x, t) = \mathcal{Q}(t)^T \tilde{f}(y, t) \tag{1.4}
$$

where $\mathcal{Q}(0) = I$, and $x_C(0) = 0$ with $\mathcal{Q}$ orthogonal linear transformation

$$
\mathcal{Q}(t) \cdot \mathcal{Q}(t)^T = \mathcal{Q}^T(t) \cdot \mathcal{Q}(t) = I.
$$

---

[1] The system was proposed for the first time by French engineer C.L.M.H. Navier in 1822 on the basis
of a suitable molecular model. In this regard, we wish to quote the following comment of Truesdell
(1953): "Such models were not new, having occurred in philosophical or qualitative speculations
for millennia past. Navier's magnificent achievement was to put these notions into a sufficiently
concrete form that he could derive equations of motion for them." However, it was only later, by the
effort of Poisson (1831), de Saint Venant (1843), and, mainly by clarifying work of Stokes (1845),
the equations found a completely satisfactory justification on the basis of the continuum mechanics
approach.

Then, (u, p) satisfies – after a linearization around $u = 0$ – the system

$$\partial_t u - \nu \Delta u + \nabla p -$$
$$-\big[\big(\omega \times x + \mathcal{Q}(t)^T u_\infty\big) \cdot \nabla\big]u + \omega \times u = f \quad \text{in } \Sigma \times (0, \infty),$$
$$\div u = 0 \quad \text{in } \Sigma \times (0, \infty), \qquad (1.5)$$
$$u = u_{\partial\Sigma} \quad \text{on } \partial\Sigma \times (0, \infty),$$
$$u(x, t) \to 0 \quad \text{as } |x| \to \infty$$

in a time-independent exterior domain $\Sigma \subset \mathbb{R}^3$, $(\mathfrak{D})^c = \Sigma$, where $\omega = \tilde{\omega} = k e_3$ and $u_{\partial\Sigma}(x, t) = \omega \times x + \mathcal{Q}(t)^T u_\infty$. For details of this coordinate transform in an even more general setting leading from (1.2)–(1.5) see [31, Chap. 1].

Remark 1 Note that if u_∞ is transversal or even orthogonal to e_3, then (3.2) contains the time-dependent term $(\mathcal{Q}(t)^T u_\infty) \cdot \nabla u$ which appears in a natural way for an observer sitting on the rotating and translating obstacle and seeing the fluid flowing past him from the time-dependent direction $\mathcal{Q}(t)^T u_\infty$.

Remark 2 Note that, because of the new coordinate system attached to the rotating body, Eq. (1.5)$_1$ contains three new terms, the classical Coriolis force term $\omega \times u$ (up to a multiplicative constant) and the terms $((\omega \times x) \cdot \nabla)u$ and $(\mathcal{Q}_\omega(t)^T v_\infty \cdot \nabla)u$ which are not subordinate to the Laplacian in unbounded domains.

Chapter 2
Formulation of the Problem

We will consider the system arising from the problem when a rigid body is translating with constant velocity and is rotating at constant angular velocity in an incompressible viscous fluid. The flow field around this body is usually described by a "modified" Navier–Stokes system, which can be written in a normalized form, and this system reads

$$\partial_t u + u \cdot \nabla u - \Delta_x u - \mathcal{R}e\Big((U + \omega \times x) \cdot \nabla_x u + \omega \times u\Big) + \nabla_x p = F, \qquad (2.1)$$

$$\text{div}_x u = 0$$

$$\text{in } (\mathbb{R}^3 \backslash \overline{\mathfrak{D}}) \times (0, T),$$

where $\mathcal{R}e = d\,w\,l/\nu$ is the Reynolds number, w is a suitable scale velocity, l is a suitable scale length, d is a constant density and ν denotes the viscosity coefficient.

Here $\mathfrak{D} \subset \mathbb{R}^3$ is a bounded domain representing the rigid body. The function u denotes the dimensionless velocity of fluid with respect to a system of coordinates whose origin is located at the center of mass of the rigid body. The function p denotes the pressure in the fluid, the vector U corresponds to the translation of the body, the vector ω corresponds to the angular velocity of the body, and the function F stands for an exterior force exerted on the fluid.

Remark 3 We use same notation u, p for the normalized form of the Navier–Stokes equations.

In the work we also consider a stationary linearized variant of (2.1) given by

$$-\Delta u - (U + \omega \times x) \cdot \nabla u + \omega \times u + \nabla \pi = f, \quad \text{div } u = 0 \quad \text{in } \mathbb{R}^3 \backslash \overline{\mathfrak{D}} \qquad (2.2)$$

under the assumption that U and ω are parallel. We will consider that $\mathcal{R}e = 1$.

This condition does not imply loss of generality; see [33, Sect. 1]. Our aim is to derive a representation formula for the velocity part u of a solution (u, π) to (2.2).

© Atlantis Press and the author(s) 2016
Š. Nečasová and S. Kračmar, *Navier–Stokes Flow Around a Rotating Obstacle*, Atlantis Briefs in Differential Equations 3, DOI 10.2991/978-94-6239-231-1_2

This formula is based on a fundamental solution to (2.2) proposed by Guenther and Thomann in the article [36] where they construct the fundamental solution to a linearized version of the time-dependent problem (2.1). In [36, p. 20], they indicate that by integrating this solution with respect to time on $(0, \infty)$, a fundamental solution to (2.2) could be obtained. They left the problem unsolved. It is this time integral we will use in our representation formula, see Theorem 5.3.

2.1   Notations, Definitions and Auxiliary Results

If $x, y \in \mathbb{R}^3$, we write $x \times y$ for the usual vector product of x and y. The open ball centered at $x \in \mathbb{R}^3$ and with radius $r > 0$ is denoted by $B_r(x)$. If $x = 0$, we will write B_r instead of $B_r(0)$. The symbol $|\cdot|$ will be used to denote the Euclidean norm in $\mathbb{R}^3$ and it will also stand for the length $\alpha_1 + \alpha_2 + \alpha_3$ of a multiindex $\alpha \in \mathbb{N}_0^3$.

We fix vectors $U, \omega \in \mathbb{R}^3 \backslash \{0\}$ which are parallel. By another transformation of variables, we may suppose there is some $\tau > 0$ with $U = -\tau \cdot e_1 = -\tau \cdot (1, 0, 0)$, hence $\omega = \varrho \cdot (1, 0, 0)$ for some $\varrho \in \mathbb{R} \backslash \{0\}$. By the symbol $\mathfrak{C}$ we denote constants depending only on U and ω. We write $\mathfrak{C}(\gamma_1, \ldots, \gamma_n)$ for constants which additionally depend on quantities $\gamma_1, \ldots, \gamma_n \in \mathbb{R}$ for some $n \in \mathbb{N}$. We further fix an open bounded set $\mathfrak{D}$ in $\mathbb{R}^3$ with Lipschitz boundary $\partial \mathfrak{D}$.

Set $p' := (1 - 1/p)^{-1}$ for $p \in (1, \infty)$.

We fix parameters $\tau \in (0, \infty)$, $\varrho \in \mathbb{R} \backslash \{0\}$, and we set $\omega := \varrho e_1$ and

$$s_\tau(x) := 1 + \tau(|x| - x_1) \quad \text{for} \ x \in \mathbb{R}^3.$$

Define the matrix by

$$\Omega := \begin{pmatrix} 0 & -\omega_3 & \omega_2 \\ \omega_3 & 0 & -\omega_1 \\ -\omega_2 & \omega_1 & 0 \end{pmatrix}$$

such that $\omega \times x = \Omega \cdot x$ for $x \in \mathbb{R}^3$. For open sets $V \subset \mathbb{R}^3$, sufficiently smooth functions $w : V \mapsto \mathbb{R}^3$, and for $z \in V$, we set

$$\mathcal{L}(w)(z) := -\Delta w(z) - (U + \omega \times z) \cdot \nabla w(z) + \omega \times w(z). \tag{2.3}$$

Let $\mathfrak{D}$ be an open bounded set in $\mathbb{R}^3$ with C^2-boundary $\partial \mathfrak{D}$. This set will be kept fixed throughout. We denote its outward unit normal by $n^{(\mathfrak{D})}$. For $T \in (0, \infty)$, put $\mathfrak{D}_T := B_T \backslash \overline{\mathfrak{D}}$ "truncated exterior domain").

For $p \in [1, \infty)$, $k \in \mathbb{N}$, and for open sets $A \subset \mathbb{R}^3$, we write $W^{k,p}(A)$ for the usual Sobolev spaces of order k and exponent p. Its standard norm will be denoted by $\|\cdot\|_{k,p}$. If $B \subset \mathbb{R}^3$ is open, define $W^{k,p}_{\text{loc}}(B)$ as the set of all functions $g : B \mapsto \mathbb{R}$ such that $g|U \in W^{k,p}(U)$ for any open bounded set $U \subset \mathbb{R}^3$ with $\overline{U} \subset B$. Also we will need the fractional order Sobolev space $W^{2-1/p,p}(\partial\mathfrak{D})$ equipped with its intrinsic norm, which we denote by $\|\cdot\|_{2-1/p,p}$ $\left(p \in (1,\infty)\right)$; see [51] for the corresponding definitions. If $\mathfrak{H}$ is a normed space whose norm is denoted by $\|\cdot\|_{\mathfrak{H}}$, and if $n \in \mathbb{N}$, we equip the product space $\mathfrak{H}^n$ with a norm $\|\cdot\|_{\mathfrak{H}}^{(n)}$ defined by $\|v\|_{\mathfrak{H}}^{(n)} := \left(\sum_{j=1}^{n}\|v_j\|_{\mathfrak{H}}^2\right)^{1/2}$ for $v \in \mathfrak{H}^n$. But for simplicity, we will write $\|\cdot\|_{\mathfrak{H}}$ instead of $\|\cdot\|_{\mathfrak{H}}^{(n)}$. We denote by $\mathcal{S}(\mathbb{R}^3)$ the usual Schwartz class of test functions.

Let $z \in \mathbb{R}^3\backslash\{0\}$, We define $\mathcal{N}$ as the fundamental solution of the Poisson equation,

$$\mathcal{N}(z) = (4\pi|z|)^{-1},$$

i.e. as the kernel of the Newton potential.

For $z \in \mathbb{R}^3\backslash\{0\}$, $r \in (0,\infty)$, $\tau \in (0,\infty)$, $j,k \in \{1, 2, 3\}$, we define

$$\Psi(r) = \int_0^r (1-e^t)\, t^{-1}\, dt, \quad \Phi(z,\tau) := (4\pi\tau)^{-1}\cdot\Psi\left(\tau\cdot(|z|-z_1)/2\right),$$

$$E_{jk}(z,\tau) = \left(\delta_{jk}\,\Delta_z - \partial/\partial z_j\partial/\partial z_k\right)\Phi(z,\tau), \tag{2.4}$$

$$E_{4k}(z) = (4\cdot\pi)^{-1}z_k\,|z|^{-3}. \tag{2.5}$$

The matrix-valued function $(E_{jk})_{1\leq j\leq 4,\,1\leq k\leq 3}$ is the fundamental solution of the Oseen system $-\Delta u + \tau\,\partial_1 u + \nabla\pi = f$, $\mathrm{div}\,u = 0$ in $\mathbb{R}^3$.

By $\mathcal{R}_i$ we denote the Riesz transforms.

Chapter 3
Fundamental Solution of the Evolution Problem

3.1 Fundamental Solution of the Non-steady "Rotating" Oseen Problem

We consider the following system of equations

$$
\begin{aligned}
\partial_t u - \Delta u - \big[(U + \omega \times x) \cdot \nabla\big]u & \\
+ \, \omega \times u + \nabla p &= f \quad \text{in } \Sigma \times (0, \infty), \\
\operatorname{div} u &= 0 \quad \text{in } \Sigma \times (0, \infty), \\
u &= u_{\partial \mathfrak{D}} \quad \text{on } \partial\Sigma \times (0, \infty), \\
u(x, t) &\to 0 \quad \text{as } |x| \to \infty,
\end{aligned}
\tag{3.1}
$$

in a time-independent exterior domain $\mathfrak{D}^c = \Sigma \subset \mathbb{R}^3$.

We define the Oseen-type operator

$$
\mathcal{L}v = \mathcal{L}_{y,t}v = -\Delta v - (U + \omega \times y) \cdot \nabla v + \omega \times v .
$$

Then the fundamental tensor of (3.1) comprises a (3×3)-matrix of distributions $\Gamma(y, z, t)$ and a three-dimensional vector of distributions $Q(y, z, t)$ such that for any vector $a \in \mathbb{R}^3$ the distributions

$$
\begin{aligned}
v_z(y, t) &= \Gamma(y, z, t)a, \; t \geq s, \quad v_z(y, t) = 0, \; t < s, \\
\pi_z(y, t) &= Q(y, z, t)a, \; t \geq s, \quad \pi_z(y, t) = 0, \; t < s,
\end{aligned}
$$

solve the system

$$
\begin{aligned}
\frac{\partial v_{z,s}}{\partial t} + \mathcal{L}v_{z,s} + \nabla\pi_{z,s} &= \delta_s(t)\delta_z(y)a, \\
\operatorname{div} v_{z,s} &= 0,
\end{aligned}
\tag{3.2}
$$

© Atlantis Press and the author(s) 2016

Š. Nečasová and S. Kračmar, *Navier–Stokes Flow Around a Rotating Obstacle*,
Atlantis Briefs in Differential Equations 3, DOI 10.2991/978-94-6239-231-1_3

in the sense of distributions. Therefore for all test functions $\varphi \in C_0^\infty(\mathbb{R}^3 \times \mathbb{R})^3$

$$\langle \partial_t v_{z,s} + \mathcal{L}v_{z,s} + \nabla \pi_{z,s}, \varphi \rangle = \varphi(z,s) \cdot a = \langle \delta_z \otimes \delta_s, \varphi \cdot a \rangle$$

and div $v_{z,s} = 0$ for all $t \geq s$. Here $\delta_s(t), \delta_z(y)$ denote the point masses concentrated at $t = s$, $y = z$.

We will recall the main theorems from the work of Guenther and Thomann (see [36]) and give the main line of the proof.

Let K denote the usual fundamental solution to the heat equation, that is,

$$K(z,t) := (4 \cdot \pi \cdot t)^{-3/2} \cdot e^{-|z|^2/(4 \cdot t)} \quad \text{for } z \in \mathbb{R}^3, \ t \in (0, \infty).$$

The ensuing estimate of the heat singularity K was proved in [56].

Lemma 3.1 *For $\alpha \in \mathbb{N}_0^3$, $l \in \mathbb{N}_0$, there is $C(l, \alpha) > 0$ such that*

$$\left| \partial_x^\alpha \partial_t^l K(x,t) \right| \leq C(l, \alpha) \cdot (|x|^2 + t)^{-3/2 - |\alpha|/2 - l} \quad \text{for } x \in \mathbb{R}^3, \ t \in (0, \infty).$$

An easy computation involving the relation $\Gamma(1/2) = \sqrt{\pi}$ (letter Γ denotes here the gamma function) yields

Lemma 3.2 $\int_0^\infty K(x,t)\, dt = (4 \cdot \pi \cdot |x|)^{-1}$ *for $x \in \mathbb{R}^3 \backslash \{0\}$.*

Theorem 3.1 *The fundamental tensor $\Gamma(y,z,t)$, $Q(y,z,t)$ of the linearized problem (3.1) can be written in the form (where $z(t) = e^{-t\Omega}z - tU$)*

$$\Gamma(y,z,t) = K(y - z(t), \tau) \left\{ \left[I - \frac{(y - z(t)) \otimes (y - z(t))}{|y - z(t)|^2} \right] \right.$$
$$\left. - {}_1F_1\left(1, \frac{5}{2}, \frac{|y - z(t)|^2}{4\tau}\right) \left[\frac{1}{3}I - \frac{(y - z(t)) \otimes (y - z(t))}{|y - z(t)|^2} \right] \right\} e^{-t\Omega},$$

$$Q(y,z,t) = -\frac{1}{4\pi} \nabla_y \frac{1}{|y - z(t)|} \delta_0 \equiv Q^*(y)\delta_0.$$

In particular, for every initial value $u_0 \in \mathcal{S}(\mathbb{R}^3)^3$

$$\lim_{(y,t) \to (y_0, 0^+)} \int_{\mathbb{R}^3} \Gamma(y,z,t)u_0(z)\, dz = Pu_0(y), \quad y \in \mathbb{R}^3,$$

where P denotes the Helmholtz projection $\mathbb{R}^3$.

(${}_1F_1$ denotes the Kummer function which will defined later)

Here and in the rest of this work, we use the abbreviation

$$z(t) := e^{-t\cdot\Omega} \cdot z - t \cdot U = e^{-t\cdot\Omega} \cdot z + \tau \cdot t \cdot e_1 \quad \text{for } z \in \mathbb{R}^3, \ t \in [0, \infty). \quad (3.3)$$

In the following we will also use the cylindrical coordinates $r, \theta, x_3 \in [0, \infty) \times [0, 2\pi) \times \mathbb{R}$ for x such that $(\omega \times x) \cdot \nabla u = \partial_\theta u$ where ∂_θ denotes the angular derivative with respect to θ. Obviously $-\Delta$ commutes with ∂_θ. Let $\nabla' = (\partial_1, \partial_2)$.

Let us define the function space $\mathcal{J}_T^{q,s}$, $1 < q, s < \infty$, of initial values with the norm

$$\|u_0\|_{\mathcal{J}_T^{q,s}} = \left(\int_0^T \left(\|e^{-tA_q} P_q u_0\|_q^s + \|A_q e^{-tA_q} P_q u_0\|_q^s \right) dt \right)^{1/s},$$

where P_q is the Helmholtz projection on $L^q(\mathbb{R}^3)^3$ and $A_q = -P_q \Delta$ is the Stokes operator. The following theorem states that the equation under consideration is well posed in this space.

Theorem 3.2 Let $0 < T < \infty$ and assume that for some $1 < q, s < \infty$ the data $u_0 \in L_\sigma^q(\mathbb{R}^3)^3$ and $f \in L^s(0, T; L^q(\mathbb{R}^3)^3)$ satisfy

$$f, \partial_\theta f, \in L^s(0, T; L^q(\mathbb{R}^3)^3),$$

and $u_0, \partial_\theta u_0 \in \mathcal{J}_T^{q,s}$. Then the unique solution $(v, \nabla p) \in L^s(0, T; (L^q(\mathbb{R}^3))^6$ of

$$\frac{\partial v}{\partial t} + \mathcal{L}v + \nabla p = f \quad in \ \mathbb{R}^3 \times (0, \infty),$$
$$\nabla \cdot v = 0 \quad in \ \mathbb{R}^3 \times (0, \infty), \qquad (3.4)$$

with the initial data $v(0, y) = u_0(y)$ is given by

$$v(y, t) = \int_0^t \int_{\mathbb{R}^3} \Gamma(y, z, t - s) f(z, s) \, dz \, ds + \int_{\mathbb{R}^3} \Gamma(y, z, t) u_0(z) \, dz, \qquad (3.5)$$

$$p(y, t) = \int_0^t \int_{\mathbb{R}^3} Q(y, z, t) \cdot f(z, s) \, dz \, ds$$
$$= \int_{\mathbb{R}^3} Q^*(y - z) \cdot f(z, t) \, dz. \qquad (3.6)$$

Moreover, v, p satisfy the a priori estimate

$$\|v; \ \nabla v; \ \nabla^2 v; \ v_t; \ \partial_\theta v; \ \nabla p\|_{L^s(0,T;L^q)} \leq C(1 + T)\big[\|u_0; \ \partial_\theta u_0\|_{\mathcal{J}_T^{q,s}} + \ + \|f; \ \partial_\theta f\|_{L^s(0,T;L^q)}\big], \qquad (3.7)$$

where the constant C depends on q, s and ω, but not on T.

Remark 4 Terms which are already present in the estimate of v_t are due to the fact that the operator $\mathcal{L}$ does not generate an analytic semigroup and will not satisfy the standard maximal regularity estimate, see [24, 25, 35, 38].

Corollary 3.1 (i) The fundamental solution Γ from Theorem 3.1 is unique.
(ii) For any $y, z \in \mathbb{R}^3$ and $s < \tau < t$ one has the semigroup property

$$\int_{\mathbb{R}^3} \Gamma(y, z', t - \tau) \Gamma(z', z, \tau - s) \, dz' = \Gamma(y, z, t - s). \tag{3.8}$$

(iii) For $u \in \mathcal{S}(\mathbb{R}^3)^3$

$$\lim_{(y,t) \to (y^0, 0^+)} \int_{\mathbb{R}^3} \Gamma(y, z, t) u(z) \, dz = Pu(y^0).$$

(iv) The (backward in time) adjoint problem

$$(-\partial_s + \mathcal{L}^*)w + \nabla \pi = g, \quad \nabla \cdot w = 0 \quad \text{on } \mathbb{R}^3 \times (0, T), \quad w(T) = 0,$$

with the operator $\mathcal{L}^* w = -\Delta w + (U + \omega \times y) \cdot \nabla w - \omega \times w$ has the fundamental solution

$$\Gamma'(z, y, s) = \Gamma(y, z, t).$$

Let P denote the Helmholtz projection of vector fields in $\mathbb{R}^3$ onto divergence free vector fields. Then,

$$P = I + \mathcal{R} = I + \nabla \text{div}(-\Delta)^{-1},$$

where $\mathcal{R}$ is the (3×3)-matrix operator with entries $(\mathcal{R}_i \mathcal{R}_j)_{i,j}$.

The Kummer function $_1F_1(1, c, u)$ appearing in the following is defined by

$$_1F_1(1, c, u) := \sum_{n=0}^{\infty} \left(\Gamma(c) / \Gamma(n + c) \right) \cdot u^n \quad \text{for } u \in \mathbb{R}, \ c \in (0, \infty),$$

where the letter Γ denotes the usual gamma function. As in [36], the same letter Γ is used to denote the fundamental solution introduced in that latter reference for a linearized version of (2.1).

As basic results for Kummer functions we mention the following facts:

Lemma 3.3 *For a, $c > 0$ the following results hold:*
(1)

$$_1F_1(1, c, \lambda) = \sum_{n=0}^{\infty} \frac{1}{(c)_n} \lambda^n.$$

(2)

$$\frac{d}{d\lambda} \, _1F_1(a, c, \lambda) = \frac{a}{c} \, _1F_1(a + 1, c + 1, \lambda).$$

(3) There exists a constant $C > 0$ such that for all $\lambda > 0$

$$\left| e^{-\lambda} \left(_1F_1(1, c, \lambda) - 1 \right) \right| \leq C \frac{\lambda}{(1 + \lambda)^c}.$$

(4)

$$\frac{d}{d\lambda}\left(e^{-\lambda}\left({}_1F_1(1, c, \lambda) - 1\right)\right) = \frac{1}{c}e^{-\lambda}{}_1F_1(1, c+1, \lambda) - \frac{\lambda}{c+1}e^{-\lambda}{}_1F_1(1, c+2, \lambda),$$

$$\frac{d^2}{d\lambda^2}\left(e^{-\lambda}\left({}_1F_1(1, c, \lambda) - 1\right)\right) = \frac{-2}{c+1}e^{-\lambda}{}_1F_1(1, c+2, \lambda) + \frac{\lambda}{c+2}e^{-\lambda}{}_1F_1(1, c+3, \lambda).$$

In particular, there exists a constant $C > 0$ such that for all $\lambda > 0$ and for $j = 1, 2$

$$\left|\frac{d^j}{d\lambda^j}\left(e^{-\lambda}\left({}_1F_1(1, c, \lambda) - 1\right)\right)\right| \leq C\frac{1}{(1+\lambda)^{c+j-1}}.$$

Proof (1)–(2) can be found in [36, pp. 82]. For the proof of (3) we use the gamma function Γ, the asymptotic result

$$e^{-\lambda}{}_1F_1(1, c, \lambda) \sim \Gamma(c)\frac{1}{\lambda^{c-1}} \quad \text{as} \quad \lambda \to \infty, \qquad\qquad . \quad (3.9)$$

see [36, p. 82], and that ${}_1F_1(1, c, 0) = 1$. (4)$_1$ follows from the formula

$$\frac{d}{d\lambda}\left(e^{-\lambda}{}_1F_1(1, c, \lambda)\right) = \frac{1-c}{c}e^{-\lambda}{}_1F_1(1, c+1, \lambda),$$

see [36, Lemma 2.1], and the identity

$${}_1F_1(1, c, \lambda) - 1 = \frac{1}{c}\lambda{}_1F_1(1, c+1, \lambda),$$

see [36, (4.9)]. The second equation in (4) is proved analogously. The estimates follows from (3.9). ∎

First, ignoring the pressure term and the solenoidality condition in (2.2), we consider the linear operator

$$\tilde{\mathcal{L}}w = \tilde{\mathcal{L}}_{y,t}w = -\Delta w - (U + \omega \times y) \cdot \nabla w + \omega \times w. \qquad (3.10)$$

Proposition 3.1 *Assume $w_0 \in \mathcal{S}(\mathbb{R}^3)^3$. Then the solution of the initial value problem*

$$\frac{\partial w}{\partial t} + \tilde{\mathcal{L}}w = 0 \quad \text{in } (0, \infty), \quad w(\cdot, s) = w_0, \qquad (3.11)$$

is given by

$$w(y, t) = \int_{\mathbb{R}^3} \tilde{\Gamma}(y, z, t)w_0(z)\, dz, \qquad (3.12)$$

where

$$\widetilde{\Gamma}(y, z, t) = K(y - z(t), t)e^{-t\Omega}. \tag{3.13}$$

Proof By two elementary transformations we will reduce the problem (3.11) to the simpler problem

$$\frac{\partial v}{\partial t} - (\omega \times y) \cdot \nabla v - \Delta v = 0 \quad \text{in } (0, \infty), \ v(0) = w_0. \tag{3.14}$$

First let $w^*(t) = \exp(t\Omega)w$. Then

$$\frac{\partial w^*}{\partial t} - (U + \omega \times y) \cdot \nabla w^* - \Delta w^* = 0 \quad \text{in } (0, \infty), \ w^*(0) = w_0. \tag{3.15}$$

Next, let

$$v(y, t) = w^*(y - Ut, t). \tag{3.16}$$

Then we are get the problem

$$\frac{\partial v}{\partial t} = (\omega \times y) \cdot \nabla v + \Delta v. \tag{3.17}$$

Now the solution v to (3.17) can be written in the form

$$\begin{aligned}
v(y, t) &= \int_{\mathbb{R}^3} K(z, t) \, w_0 \left(e^{t\Omega}y - z \right) dz \\
&= \frac{1}{(4\pi t)^{3/2}} \int_{\mathbb{R}^3} \exp \left(-\frac{|e^{t\Omega}y - z|^2}{4t} \right) w_0(z) \, dz \,,
\end{aligned}$$

see e.g. DaPrato and Lunardi [49]. Hence we get

$$\begin{aligned}
w(y, t) &= e^{-t\Omega} w^*(y, t) = e^{-t\Omega} v(y + tU, t) \\
&= \frac{1}{(4\pi t)^{3/2}} e^{-t\Omega} \int_{\mathbb{R}^3} \exp \left(-\frac{|e^{t\Omega}(y + tU) - z|^2}{4t} \right) w_0(z) \, dz \,. \qquad \blacksquare
\end{aligned}$$

To obtain the fundamental solution of the linearized problem (3.1) taking into account the incompressibility condition, we have to adapt Proposition 3.1, cf. [36]. Using the Helmholtz projection P it is easy to see that for every fixed $a \in \mathbb{R}^3$

$$\begin{aligned}
\Gamma(y, z, t)a &= P(\widetilde{\Gamma}(y, z, t)a), \\
Q(y, z, t)a &= -\tfrac{1}{4\pi} a \cdot \nabla \tfrac{1}{|y-z|} \delta_0(t),
\end{aligned}$$

is the fundamental tensor for the linear equation (3.1); here, P acts on the variable y. In particular, for $t > 0$

$$\left(\frac{\partial}{\partial t} + \mathcal{L}\right)(\Gamma a) + \nabla Q a = 0, \quad \nabla \cdot (\Gamma a) = 0. \tag{3.18}$$

Since

$$\Gamma(y, z, t)a = P\widetilde{\Gamma}(y, z, t)a = (I + \mathcal{R})\widetilde{\Gamma}(y, z, t)a = [(I + \mathcal{R})K(y - z(t), t]e^{-t\Omega}a,$$

and $\mathcal{R}_i \mathcal{R}_j f = \frac{\partial}{\partial y_i} \frac{\partial}{\partial y_j}(-\Delta)^{-1} f$, we get

$$\Gamma(y, z, t)a = [K(y - z(t), t)I + \text{Hess } \psi(y, z, \tau)]e^{-t\Omega}a, \tag{3.19}$$

here $\psi(y, z, t)$ is the solution of the equation $-\Delta_y \psi(y, t) = K(y - z(t), t)$, i.e.,

$$\psi(y, z, t) = \frac{1}{4\pi} \frac{1}{(4\pi t)^{3/2}} \int_{R^3} \frac{1}{|y - x|} \exp\left(-\frac{|x - z(t)|^2}{4t}\right) dx, \tag{3.20}$$

and Hess $\psi(y, \tau) = \left(\frac{\partial}{\partial y_i} \frac{\partial}{\partial y_j}\right)\psi(y, \tau)$ denotes the Hessian of ψ.

To compute ψ and its Hessian we follow [36] and introduce the *error function*

$$\text{Erf}(s) = \frac{2}{\sqrt{\pi}} \int_0^s e^{-u^2} du = \frac{2s}{\sqrt{\pi}} e^{-s^2} {}_1F_1(1, 3/2, s^2).$$

Lemma 3.4 *For all $t > 0$*

$$\psi(y, z, t) = \frac{1}{4\pi|y - z(t)|} \text{Erf}\left(\frac{|y - z(t)|}{\sqrt{4t}}\right) \tag{3.21}$$

and

$$\frac{\partial^2}{\partial y_i \partial y_j}\psi(y, z, t) = K(y - z(t), t)\left(-\frac{1}{3}{}_1F_1\left(1, \frac{5}{2}, \frac{|y - z(t)|^2}{4t}\right)\delta_{ij}\right.$$

$$\left. + \frac{(y_i - z_i(t))(y_j - z_j(t))}{|y - z(t)|^2}\left[{}_1F_1\left(1, \frac{5}{2}, \frac{|y - z(t)|^2}{4t}\right) - 1\right]\right).$$

Proof See [36, Lemma 3.1, Proposition 3.2].

From Proposition 3.1 and Lemma 3.4 it follows for all $a \in \mathbb{R}^3$ that $(\partial_t + \mathcal{L})(\Gamma a) + \nabla(Qa) = 0$ for $t > 0$ and div $(\Gamma a) = 0$. It remains to show, for every initial value $u_0 \in \mathcal{S}(\mathbb{R}^3)^3$ with the Helmholtz decomposition $u_0 = h + \nabla q$, that

$$\lim_{t \to 0+} \int_{\mathbb{R}^3} \Gamma(y, z, t)u_0(z)\, dz + \nabla_y \int_{\mathbb{R}^3} Q^*(y, z)u_0(z)\, dz = u_0(y). \tag{3.22}$$

It can be found in detail [36].

3.2   Basic Properties of the Fundamental Solution

We will use the following notation:

$$w = y - z(t), \quad \hat{w} = \frac{w}{|w|},$$

$$\Lambda(\hat{w}) = \hat{w} \otimes \hat{w},$$

$$\lambda = \frac{|y - z(t)|^2}{4t},$$

$$\mathcal{F}(\lambda) = {}_1F_1(1, 5/2, \lambda),$$

$$M(y, z, t, s) = \frac{1}{3} \frac{1}{(4\pi t)^{3/2}} e^{-\lambda} \mathcal{F}(\lambda)[I - 3\Lambda(\hat{w})],$$

so that

$$\Gamma(y, z, t) = \big[K(y - \tilde{z}(t), t)\{I - \Lambda(\hat{w})\} - M(y, z, t)\big] e^{-t\Omega}.$$

Proposition 3.2 *The fundamental solution Γ has (in each component of the (3×3)-matrix) the following asymptotic properties:*

(i) For any vectors $y, z \in \mathbb{R}^3$, $y \neq z$,

$$\Gamma(y, z, t) \sim -\frac{1}{4\pi} \frac{1}{|y - z|^3} \left[I - 3\frac{(y - z) \otimes (y - z)}{|y - z|^2}\right] \quad as \ t \to 0.$$

(ii) For any vectors $y, z \in \mathbb{R}^3$ and $t \to \infty$,

- $U = 0$

$$e^{t\Omega} \Gamma(y, z, t) \sim \frac{2}{3} \frac{1}{(4\pi(t))^{3/2}} I,$$

- $U \neq 0$

$$\Gamma(y, z, t) \sim -\frac{1}{4\pi} \frac{1}{|tU|^3} \left[I - 3\frac{U \otimes U}{|U|^2}\right].$$

(iii) Let $y^0, z, s \in \mathbb{R}^3$, $|s| = 1$, be fixed and let $y = y^0 + \rho\zeta$, $\rho > 0$. Then for $t > 0$

$$\Gamma(y, z, t) \sim -\frac{1}{4\pi} \frac{1}{|y - z(t)|^3}[I - 3\zeta \otimes \zeta] \quad as \ \rho \to \infty.$$

Proof (i) Since $y \neq z$, the term $\lambda \to \infty$ as $t \to 0$. Hence the leading term in Γ is determined by M, where by Lemma 3.3(3), $e^{-\lambda}\mathcal{F}(\lambda) \sim \Gamma(5/2)\lambda^{-3/2} = \frac{3}{4}\sqrt{\pi}\lambda^{-3/2}$. This proves (i).
(iii) In this case $\lambda \to \infty$ and the leading term in Γ is determined by M, cf. (i). Since $\Lambda(\hat{w}) \sim \zeta \otimes \zeta$ as $\rho \to \infty$ for $t > s$ fixed, we get (iii).

(ii) In the case $U = 0$ then $u = |y - e^{-t\Omega}z|^2/4t \to 0$ and

$$e^{-u}[1 - {}_1F_1(1, 5/2, u)] \to 0,$$

therefore the first part of (ii) is proven.

In the second case $\tilde{z}(t) = e^{-t\Omega}z - tU$ and we get for large t that

$$\lambda = \frac{|y - z(t)|^2}{4t} \sim \frac{t|U|^2}{4},$$

$$\hat{w} = \frac{y - z(t)}{|y - z(t)|} \sim \frac{U}{|U|},$$

Since by Lemma 3.3 (3) the leading term in Γ is determined by M, we have

$$e^{t\Omega}\Gamma(y, z, t) \sim -\frac{1}{3}\frac{\Gamma(5/2)}{(4\pi t)^{3/2}}\left(\frac{4}{|tU|^2}\right)^{3/2}\left[I - 3\frac{U \otimes U}{|U|^2}\right]$$

$$= -\frac{1}{4\pi}\frac{1}{|tU|^3}\left[I - 3\frac{U \otimes U}{|U|^2}\right]$$

as $t \to \infty$.

∎

We will give summary of previous results Proposition 3.2 and Theorems 3.1, 3.2 with results from [36]:

Theorem 3.3 ([36, Proposition 4.1], [Proposition 3.2(i)]) *Let* $j, k \in \{1, 2, 3\}$, $y, z \in \mathbb{R}^3$ *with* $y \neq z$. *Then*

$$\Gamma_{jk}(y, z, t) \to -(4 \cdot \pi)^{-1} \cdot \left(\delta_{jk} \cdot |y - z|^{-3} - 3 \cdot (y - z)_j \cdot (y - z)_k \cdot |y - z|^{-5}\right)$$

for $t \downarrow 0$.

Theorem 3.4 ([36, Theorem 1.3], Theorems 3.1, 3.2) *Let* $y, z \in \mathbb{R}^3$ *with* $y \neq z$, $t \in (0, \infty)$, $j, k \in \{1, 2, 3\}$. *Then*

$$\partial_t\Gamma_{jk}(y, z, t) - \Delta_z\Gamma_{jk}(y, z, t) + (U + \omega \times z) \cdot \nabla_z\Gamma_{jk}(y, z, t)$$

$$-\left[\omega \times \left(\Gamma_{js}(y, z, t)\right)_{1 \leq s \leq 3}\right]_k = 0.$$

We further set

$$E_{4j}(x) := (4\pi)^{-1}x_j|x|^{-3} \quad (1 \leq j \leq 3, \ x \in \mathbb{R}^3\backslash\{0\}).$$

E_{4j} is the pressure part of the fundamental tensor of the stationary Oseen problem. It follows:

$$Q(y, z, t) = E_{4j}(y - z(t))\delta_0(t) = \nabla\mathcal{N}(y - z(t))\delta_0(t).$$

Theorem 3.5 ([36, Proposition 4.1], Theorems 3.1, 3.2) *Let $j, k \in \{1, 2, 3\}$, $y, z \in \mathbb{R}^3$. Then*

$$\Gamma_{jk}(y, z, t) \to -\partial_k E_{4j}(y - z) \quad for \ t \downarrow 0 \ \ if \ y \neq z. \tag{3.23}$$

3.3   Further Properties of Γ_{jk}

We derive some additional properties of the fundamental solution of the evolution problem in order to get the fundamental solution of the stationary problem.

3.3.1   Technical Lemmas

Concerning the term $s_\tau(x)$, we will need the following estimates.

Lemma 3.5 ([15, Lemma 4.3]) *Let $\beta \in (1, \infty)$. Then*

$$\int_{\partial B_r} s_\tau(x)^{-\beta} \, do_x \leq \mathfrak{C}(\beta) r \ \ for \ r \in (0, \infty).$$

Lemma 3.6 ([7, Lemma 4.8]) *For $x, y \in \mathbb{R}^3$, we have*

$$s_\tau(x - y)^{-1} \ \leq \ \mathfrak{C}(1 + |y|) s_\tau(x)^{-1}.$$

Lemma 3.7 ([4, Lemma 2]) *Let $S \in (0, \infty)$, $x \in B_S$, $t \in (0, \infty)$. Then*

$$|x - \tau t e_1|^2 + t \geq \mathfrak{C}(S) (|x|^2 + t).$$

Lemma 3.8 *Let $S \in (0, \infty)$, $x \in B_S^c$. Then $|x| \geq \mathfrak{C}(S) s_\tau(x)$.*

Proof $|x| \geq S/2 + |x|/2 \geq S/2 + (|x| - x_1)/4 \geq \min\{S/2, 1/(4\tau)\} s_\tau(x)$. ∎

3.3.2   Pointwise Estimates of Γ_{jk}

We define some further functions which will be useful in the following:

$$\mathfrak{S}(u) := \sum_{n=1}^{\infty} \Gamma(5/2) \cdot \Gamma(5/2+n)^{-1} \cdot u^{n-1} \quad \text{for} \ u \in \mathbb{R},$$

$$\Lambda_{jk}(x,t) := K(x,t) \cdot \left(\delta_{jk} - x_j \cdot x_k \cdot |x|^{-2} \right.$$
$$\left. - {}_1F_1\left(1, 5/2, |x|^2/(4 \cdot t)\right) \cdot \left(\delta_{jk}/3 - x_j \cdot x_k \cdot x^{-2} \right) \right)$$

for $x \in \mathbb{R}^3 \backslash \{0\}$, $t \in (0, \infty)$, $j, k \in \{1, 2, 3\}$. Note that

$$\left(\Gamma_{jk}(y, z, t) \right)_{1 \le j, k \le 3} = \left(\Lambda_{rs}(y - z(t), t) \right)_{1 \le r, s \le 3} \cdot e^{-t \cdot \Omega} \tag{3.24}$$

for $y, z \in \mathbb{R}^3$, $t \in (0, \infty)$ with $y \ne z(t)$.

The ensuing estimates of Kummer functions are standard (see [50], (3.25)).

Theorem 3.6 *Let $S \in (0, \infty)$. Then there is $C(S) > 0$ such that*

$$\left| d^k/du^k \left(e^{-u} \cdot {}_1F_1(1, 5/2, u)^r \right) \right| \le C(S) \cdot u^{-3/2-k} \tag{3.25}$$
for $u \in [S, \infty)$, $k \in \{0, 1, 2\}$,
$$\left| d^k/du^k \, {}_1F_1(1, 5/2, u) \right| + \left| \mathfrak{S}^{(k)}(u) \right| \le C(S), \quad \left| -{}_1F_1(1, 5/2, u) + 1 \right| \le C(S) \cdot |u|$$
for $u \in [-S, S]$, $k \in \{0, 1, 2\}$.

Of course, in the preceding theorem one might admit any $k \in \mathbb{N}$, but we will need only values k from $\{0, 1, 2\}$.

The ensuing lemma is easy to verify.

Lemma 3.9 *For $y, z \in \mathbb{R}^3$, $t \in (0, \infty)$, put*

$$\tilde{\Gamma}(y, z, t) := K(y - z(t), t) \cdot \left[\left(1 - {}_1F_1\left(1, 5/2, |y - z(t)|^2/(4 \cdot t)\right)/3 \right) \cdot \delta_{rs} \right.$$
$$\left. + \mathfrak{S}\left(|y - z(t)|^2/(4 \cdot t) \right) \cdot (4 \cdot t)^{-1} \cdot \left(y - z(t) \right)_r \cdot \left(y - z(t) \right)_s \right]_{1 \le r, s \le 3} \cdot e^{-t \cdot \Omega},$$

where $z(t)$ was introduced in (3.3). Then

$$\Gamma_{jk}(y, z, t) = \tilde{\Gamma}_{jk}(y, z, t) \quad \text{for} \ y, z \in \mathbb{R}^3, \ t \in (0, \infty) \ \text{with} \ y \ne z(t). \tag{3.26}$$

Since $\tilde{\Gamma}_{jk} \in C^{\infty}\left(\mathbb{R}^3 \times \mathbb{R}^3 \times (0, \infty) \right)$, we may conclude

Corollary 3.2 *The function Γ_{jk} may be extended continuously to a function from $C^{\infty}\left(\mathbb{R}^3 \times \mathbb{R}^3 \times (0, \infty) \right)$.*

Next we state that the function $\Gamma_{jk}(y, z, t)$ and its first order derivatives with respect to z may be estimated in a similar way as $K(y - z(t), t)$ and $\partial_{y_l} K(y - z(t), t)$, respectively ($1 \le l \le 3$).

Lemma 3.10 *For $x \in \mathbb{R}^3$, $t \in (0, \infty)$, $\beta \in \mathbb{N}_0^3$ with $|\beta| \leq 1$, the inequality*

$$|\partial_x^\beta \Lambda_{jk}(x, t)| \leq \mathfrak{C} \cdot (|x|^2 + t)^{-3/2 - |\beta|/2}$$

holds. By (3.24), this means that

$$|\partial_y^\beta \Gamma_{jk}(y, z, t)| + |\partial_z^\beta \Gamma_{jk}(y, z, t)| \leq \mathfrak{C} \cdot (|y - z(t)|^2 + t)^{-3/2 - |\beta|/2}$$

for $y, z \in \mathbb{R}^3$ and for t, β as above.

Proof Take x, t as in the lemma and suppose in addition that $x \neq 0$. Let $n \in \{1, 2, 3\}$. Then we distinguish two cases. First we suppose that $|x|^2 \leq t$. It follows from (3.26), Lemma 3.1 and Theorem 3.6 with $s = 1/4$ that

$$|\Lambda_{jk}(x, t)| \tag{3.27}$$
$$\leq \mathfrak{C} \cdot K(x, t) \cdot \left(1 + |{}_1F_1(1, 5/2, |x|^2/(4 \cdot t))| + |\mathfrak{S}(|x|^2/(4 \cdot t))| \cdot |x|^2/t\right)$$
$$\leq \mathfrak{C} \cdot K(x, t) \leq \mathfrak{C} \cdot (|x|^2 + t)^{-3/2},$$

$$|\partial_{x_n} \Lambda_{jk}(x, t)| \tag{3.28}$$
$$\leq \mathfrak{C} \cdot \sum_{l=1}^3 \left[|\partial_{x_l} K(x, t)| \cdot \left(1 + |{}_1F_1(1, 5/2, |x|^2/(4 \cdot t))|\right)\right.$$
$$+ K(x, t) \cdot |d/du \, {}_1F_1(1, 5/2, u)|_{u=|x|^2/(4 \cdot t)} \cdot |x|/t$$
$$+ |\partial_{x_l} K(x, t)| \cdot |\mathfrak{S}(|x|^2/(4 \cdot t))| \cdot |x|^2/t$$
$$\left. + K(x, t) \cdot |\mathfrak{S}'(|x|^2/(4 \cdot t))| \cdot |x|^3/t^2 + K(x, t) \cdot |\mathfrak{S}(|x|^2/(4 \cdot t))| \cdot |x|/t\right]$$
$$\leq \mathfrak{C} \cdot \left[(|x|^2 + t)^{-2} + (|x|^2 + t)^{-3/2} \cdot |x|/t + (|x|^2 + t)^{-2} \cdot |x|^2/t\right.$$
$$\left. + (|x|^2 + t)^{-3/2} \cdot |x|^3/t^2 + (|x|^2 + t)^{-3/2} \cdot |x|/t\right]$$
$$\leq \mathfrak{C} \cdot \left((|x|^2 + t)^{-2} + (|x|^2 + t)^{-3/2} \cdot t^{-1/2}\right) \leq \mathfrak{C} \cdot (|x|^2 + t)^{-2}.$$

Now consider the case $|x|^2 \geq t$. By Lemma 3.1 and (3.25) with $S = 1/4$,

$$|\Lambda_{jk}(x, t)| \tag{3.29}$$
$$\leq \mathfrak{C} \cdot \left(K(x, t) + t^{-3/2} \cdot |e^{-u} \cdot {}_1F_1(1, 5/2, u)|_{u=|x|^2/(4 \cdot t)}\right)$$
$$\leq \mathfrak{C} \cdot \left((|x|^2 + t)^{-3/2} + t^{-3/2} \cdot (t/|x|^2)^{3/2}\right)$$
$$\leq \mathfrak{C} \cdot \left((|x|^2 + t)^{-3/2} + |x|^{-3}\right) \leq \mathfrak{C} \cdot (|x|^2 + t)^{-3/2},$$

$$|\partial_{x_n} \Lambda_{jk}(x, t)| \tag{3.30}$$

$$\le \mathfrak{C} \cdot \sum_{l=1}^{3} \Big[|\partial_{x_l} K(x, t)| + K(x, t) \cdot |x|^{-1}$$

$$+ t^{-3/2} \cdot |d/du \left(e^{-u} \cdot {}_1F_1(1, 5/2, u) \right)|_{u=|x|^2/(4 \cdot t)} \cdot |x|/t$$

$$+ t^{-3/2} \cdot |e^{-u} \cdot {}_1F_1(1, 5/2, u)|_{u=|x|^2/(4 \cdot t)} \cdot |x|^{-1} \Big]$$

$$\le \mathfrak{C} \cdot \Big[(|x|^2 + t)^{-2} + (|x|^2 + t)^{-3/2} \cdot |x|^{-1}$$

$$+ t^{-5/2} \cdot (t/|x|^2)^{5/2} \cdot |x| + t^{-3/2} \cdot (t/|x|^2)^{3/2} \cdot |x|^{-1} \Big]$$

$$\le \mathfrak{C} \cdot \Big((|x|^2 + t)^{-2} + (|x|^2 + t)^{-3/2} \cdot |x|^{-1} + |x|^{-4} \Big) \le \mathfrak{C} \cdot (|x|^2 + t)^{-2}.$$

The lemma follows from (3.27)–(3.30). ∎

Concerning the matrix Ω, we observe

Lemma 3.11 *Let* $x \in \mathbb{R}^3$, $t \in \mathbb{R}$. *Then*

$$(e^{t\Omega} \cdot x)_1 = x_1, \quad e^{t\Omega} \cdot e_1 = e_1.$$

Proof The assertion immediately follows from the relation $\Omega = \varrho \begin{pmatrix} 0 & 0 & 0 \\ 0 & 0 & -1 \\ 0 & 1 & 0 \end{pmatrix}$. ∎

Due to Lemma 3.11, we get

Lemma 3.12

$$\left(\Gamma_{jk}(y, z, t) \right)_{1 \le j,k \le 3} = e^{-t\Omega} \cdot \left(\Lambda_{rs}(e^{t\Omega} \cdot y - \tau t e_1 - z, t) \right)_{1 \le r,s \le 3} \tag{3.31}$$

for $y, z \in \mathbb{R}^3$, $t \in (0, \infty)$.

Finally we prove an analogue of Lemma 3.10 for the second derivatives of Λ_{jk}.
We put

$$\eta_{jk}(x) := x_j \cdot x_k \cdot |x|^{-2} \quad \text{and} \quad \eta(x) := x \otimes x \cdot |x|^{-2} \quad \text{for } x \in \mathbb{R}^3 \backslash \{0\},$$

$$\Lambda_{jk}(x, t) := K(x, t) \cdot \Big(\delta_{jk} - \eta_{jk}(x) - {}_1F_1\big(1, 5/2, |x|^2/(4 \cdot t)\big) \cdot \big(\delta_{jk}/3 - \eta_{jk}(x) \big) \Big)$$

for $x \in \mathbb{R}^3 \backslash \{0\}$, $t \in (0, \infty)$, $j, k \in \{1, 2, 3\}$,

Lemma 3.13 *Let* $j, k, l, m \in \{1, 2, 3\}$, $x \in \mathbb{R}^3$, $t \in (0, \infty)$. *Then*

$$|\partial_{x_l} \partial_{x_m} \Lambda_{jk}(x, t)| \le \mathfrak{C} \cdot (|x|^2 + t)^{-5/2}.$$

Proof In view of Lemma 3.8, we may suppose $x \neq 0$. We have

$$|\partial^\beta \eta_{jk}(x)| \leq \mathfrak{C} \cdot |x|^{-|\beta|} \quad (\beta \in \mathbb{N}_0^3 \text{ with } |\beta| \leq 2). \tag{3.32}$$

Moreover, if $G \in C^2(\mathbb{R})$, the relations

$$\partial_{x_r}\left(G\left(|x|^2/(4 \cdot t)\right)\right) = G'\left(|x|^2/(4 \cdot t)\right) \cdot x_r/(2 \cdot t)$$

$$\partial_{x_l}\partial_{x_m}\left(G\left(|x|^2/(4 \cdot t)\right)\right)$$
$$= G'\left(|x|^2/(4 \cdot t)\right) \cdot \delta_{lm}/(2 \cdot t) + G''\left(|x|^2/(4 \cdot t)\right) \cdot x_l \cdot x_m/(4 \cdot t^2)$$

hold ($l, m, r \in \{1, 2, 3\}$), so that

$$\left|\partial_{x_r}\left(G\left(|x|^2/(4 \cdot t)\right)\right)\right| \leq \mathfrak{C} \cdot \left|G'\left(|x|^2/(4 \cdot t)\right)\right| \cdot (|x|^2/t)^{1/2} \cdot t^{-1/2} \tag{3.33}$$

$$\left|\partial_{x_l}\partial_{x_m}\left(G\left(|x|^2/(4 \cdot t)\right)\right)\right| \tag{3.34}$$
$$\leq \mathfrak{C} \cdot \left(\left|G'\left(|x|^2/(4 \cdot t)\right)\right| + \left|G''\left(|x|^2/(4 \cdot t)\right)\right| \cdot |x|^2/t\right) \cdot t^{-1}.$$

Now we distinguish the cases $|x|^2 \leq t$ and $|x|^2 > t$. First, suppose that $|x|^2 \leq t$. Then $|x|^2/(4 \cdot t) \leq 1/4$, $t = t/2 + t/2 \geq (|x|^2 + t)/2$, so that for $G \in C^2(\mathbb{R})$, with Lemma 3.1, (3.33) and (3.34),

$$\left|\partial_{x_r}\left(K(x, t) \cdot G\left(|x|^2/(4 \cdot t)\right)\right)\right| \tag{3.35}$$
$$\leq \left|\partial_{x_r}K(x, t) \cdot G\left(|x|^2/(4 \cdot t)\right)\right| + K(x, t) \cdot \left|\partial_{x_r}\left(G\left(|x|^2/(4 \cdot t)\right)\right)\right|$$
$$\leq \mathfrak{C} \cdot \left((|x|^2 + t)^{-2} \cdot \|G\|[0, 1/4]\|_\infty + (|x|^2 + t)^{-3/2} \cdot \|G'\|[0, 1/4]\|_\infty \cdot (|x|^2/t) \cdot t^{-1/2}\right)$$
$$\leq \mathfrak{C} \cdot (|x|^2 + t)^{-2} \cdot \sum_{\nu=0}^{1} \|G^{(\nu)}\|[0, 1/4]\|_\infty,$$

$$\left|\partial_{x_l}\partial_{x_m}\left(K(x, t) \cdot G\left(|x|^2/(4 \cdot t)\right)\right)\right| \leq \left|\partial_{x_l}\partial_{x_m}K(x, t) \cdot G\left(|x|^2/(4 \cdot t)\right)\right| \tag{3.36}$$
$$+ \sum_{(r,s)\in\{(l,m),\,(m,l)\}} \left|\partial_{x_r}K(x, t) \cdot \partial_{x_s}\left(G\left(|x|^2/(4 \cdot t)\right)\right)\right|$$
$$+ K(x, t) \cdot \left|\partial_{x_l}\partial_{x_m}\left(G\left(|x|^2/(4 \cdot t)\right)\right)\right|$$
$$\leq \mathfrak{C} \cdot \left((|x|^2 + t)^{-5/2} \cdot \|G\|[0, 1/4]\|_\infty + (|x|^2 + t)^{-2} \cdot \|G'\|[0, 1/4]\|_\infty \cdot (|x|^2/t)^{1/2} \cdot t^{-1/2}\right.$$
$$\left. + (|x|^2 + t)^{-3/2} \cdot t^{-1} \cdot \left(\|G'\|[0, 1/4]\|_\infty + \|G''\|[0, 1/4]\|_\infty \cdot |x|^2/t\right)\right)$$
$$\leq \mathfrak{C} \cdot (|x|^2 + t)^{-5/2} \cdot \sum_{\nu=0}^{2} \|G^{(\nu)}\|[0, 1/4]\|_\infty.$$

Abbreviate $\mathfrak{R}(u) := 1 - {}_1F_1(1, 5/2, u)/3$ for $u \in \mathbb{R}$. Then it follows by Theorem 3.6, Corollary 3.2 and Lemma 3.1, (3.35) and (3.36) that

$$
\begin{aligned}
&|\partial_{x_l} \partial_{x_m} \Lambda_{jk}(x, t)| \\
&\leq \left| \partial_{x_l} \partial_{x_m} \left(K(x, t) \cdot \mathfrak{R}\big(|x|^2/(4 \cdot t)\big) \right) \right| \\
&\quad + \left| \partial_{x_l} \partial_{x_m} \left(K(x, t) \cdot \mathfrak{S}\big(|x|^2/(4 \cdot t)\big) \right) \cdot (4 \cdot t)^{-1} \cdot x_j \cdot x_k \right| \\
&\quad + \sum_{(r,s) \in \{(l,m),\,(m,l)\}} \left| \partial_{x_r} \left(K(x, t) \cdot \mathfrak{S}\big(|x|^2/(4 \cdot t)\big) \right) \cdot (4 \cdot t)^{-1} \cdot \partial_{x_s}(x_j \cdot x_k) \right| \\
&\quad + \left| K(x, t) \cdot \mathfrak{S}\big(|x|^2/(4 \cdot t)\big) \cdot (4 \cdot t)^{-1} \cdot \partial_{x_l} \partial_{x_m}(x_j \cdot x_k) \right| \\
&\leq \mathfrak{C} \cdot \Big(\sum_{\nu=0}^{2} \|\mathfrak{R}^{(\nu)}|[0, 1/4]\|_\infty \cdot (|x|^2 + t)^{-5/2} \\
&\quad + \sum_{\nu=0}^{2} \|\mathfrak{S}^{(\nu)}|[0, 1/4]\|_\infty \cdot (|x|^2 + t)^{-5/2} \cdot |x|^2/t \\
&\quad + \sum_{\nu=0}^{1} \|\mathfrak{S}^{(\nu)}|[0, 1/4]\|_\infty \cdot (|x|^2 + t)^{-2} \cdot |x|/t + \|\mathfrak{S}|[0, 1/4]\|_\infty \cdot (|x|^2 + t)^{-3/2} \cdot t^{-1} \Big) \\
&\leq \mathfrak{C} \cdot \Big((|x|^2 + t)^{-5/2} \cdot (1 + |x|^2/t) + (|x|^2 + t)^{-2} \cdot (|x|^2/t)^{1/2} \cdot t^{-1/2} \\
&\quad\quad\quad\quad\quad\quad + (|x|^2 + t)^{-3/2} \cdot t^{-1} \Big).
\end{aligned}
$$

Using again that $t \geq (|x|^2 + t)/2$ and $|x|^2/t \leq 1/4$, we thus arrive at the inequality

$$
|\partial_{x_l} \partial_{x_m} \Lambda_{jk}(x, t)| \leq \mathfrak{C} \cdot (|x|^2 + t)^{-5/2} \quad \text{in the case } |x|^2 \leq t. \tag{3.37}
$$

Now we assume that $|x|^2 > t$, so that $|x|^2/(4 \cdot t) \geq 1/4$, $(|x|^2/t)^{-1} \leq 1$ and $|x|^2 = (|x|^2 + |x|^2)/2 \geq (|x|^2 + t)/2$. Abbreviate $\mathfrak{F}(u) := e^{-u} \cdot {}_1F_1(1, 5/2, u)$ for $u \in \mathbb{R}$. This means that

$$
K(x, t) \cdot {}_1F_1\big(1, 5/2, |x|^2/(4 \cdot t)\big) = (4 \cdot \pi \cdot t)^{-3/2} \cdot \mathfrak{F}\big(|x|^2/(4 \cdot t)\big).
$$

Hence

$$
\Lambda_{jk}(x, t) = K(x, t) \cdot \big(\delta_{jk} - \eta_{jk}(x)\big) - (4 \cdot \pi \cdot t)^{-3/2} \cdot \mathfrak{F}\big(|x|^2/(4 \cdot t)\big) \cdot \big(\delta_{jk}/3 - \eta_{jk}(x)\big).
$$

Thus, referring to (3.32)–(3.34), Theorem 3.6 and Lemma 3.1, we find

$$|\partial x_l \partial x_m \Lambda_{jk}(x, t)| \tag{3.38}$$

$$\leq |\partial x_l \partial x_m K(x, t) \cdot (\delta_{jk} - \eta_{jk}(x))| + \sum_{(r,s)\in\{(l,m),\,(m,l)\}} |\partial x_r K(x, t) \cdot \partial x_s \eta_{jk}(x)|$$

$$+ |K(x, t) \cdot \partial x_l \partial x_m \eta_{jk}(x)|$$

$$+ (4 \cdot \pi \cdot t)^{-3/2} \cdot \left|\partial x_l \partial x_m \Big(\mathfrak{F}(|x|^2/(4 \cdot t))\Big) \cdot \big(\delta_{jk}/3 - \eta_{jk}(x)\big)\right|$$

$$+ (4 \cdot \pi \cdot t)^{-3/2} \cdot \sum_{(r,s)\in\{(l,m),\,(m,l)\}} \left|\partial x_r \Big(\mathfrak{F}(|x|^2/(4 \cdot t))\Big) \cdot \partial x_s \eta_{jk}(x)\right|$$

$$+ (4 \cdot \pi \cdot t)^{-3/2} \cdot \left|\mathfrak{F}(|x|^2/(4 \cdot t)) \cdot \partial x_l \partial x_m \eta_{jk}(x)\right|$$

$$\leq \mathfrak{C} \cdot \Big((|x|^2 + t)^{-5/2} + (|x|^2 + t)^{-2} \cdot |x|^{-1} + (|x|^2 + t)^{-3/2} \cdot |x|^{-2}$$

$$+ t^{-5/2} \cdot \big[\,|\mathfrak{F}'(|x|^2/(4 \cdot t))| + |\mathfrak{F}''(|x|^2/(4 \cdot t))| \cdot |x|^2/t\,\big]$$

$$+ t^{-2} \cdot |\mathfrak{F}'(|x|^2/(4 \cdot t))| \cdot (|x|^2/t)^{1/2} \cdot |x|^{-1} + t^{-3/2} \cdot |\mathfrak{F}(|x|^2/(4 \cdot t))| \cdot |x|^{-2}\Big)$$

$$\leq \mathfrak{C} \cdot \Big((|x|^2 + t)^{-5/2} + \sum_{\nu=0}^{2} \sup\{\mathfrak{F}^{(\nu)}(u) \cdot u^{3/2+\nu} \,:\, u \in [1/4, \infty)\}$$

$$\cdot \big[\,t^{-5/2} \cdot (|x|^2/t)^{-5/2} + t^{-2} \cdot (|x|^2/t)^{-2} \cdot |x|^{-1} + t^{-3/2} \cdot (|x|^2/t)^{-3/2} \cdot |x|^{-2}\,\big]\Big)$$

$$\leq \mathfrak{C} \cdot \big((|x|^2 + t)^{-5/2} + |x|^{-5}\big) \leq \mathfrak{C} \cdot (|x|^2 + t)^{-5/2}.$$

The lemma follows from (3.37) and (3.38)   ∎

The preceding lemma and Lemma 3.10 yield a preliminary estimate of Γ_{jk}:

Corollary 3.3 Let $j, k \in \{1, 2, 3\}$, $\alpha, \beta \in \mathbb{N}_0^3$ with $|\alpha + \beta| \leq 2$, $y, z \in \mathbb{R}^3$, $t \in (0, \infty)$. Then

$$|\partial_y^\alpha \partial_z^\beta \Gamma_{jk}(y, z, t)| \leq \mathfrak{C} \cdot (|y - z(t)|^2 + t)^{-3/2 - |\alpha+\beta|/2}.$$

Proof By the definition of Γ_{jk}, we have

$$\Gamma_{jk}(y, z, t) = \sum_{s=1}^{3} \Lambda_{js}(y - z(t), t) \cdot (e^{-t \cdot \Omega})_{sk}.$$

Differentiating the preceding equation and applying Lemmas 3.10 and 3.13 yield Corollary 3.3.   ∎

Chapter 4
Fundamental Solution of the Stationary Problem

We consider the following system of equations

$$-\nu \Delta u - \left[(U + \omega \times x) \cdot \nabla\right]u$$
$$+\omega \times u + \nabla p = f \quad \text{in } \Sigma \times (0, \infty),$$
$$\text{div } u = 0 \quad \text{in } \Sigma \times (0, \infty), \qquad (4.1)$$
$$u = u_{\partial\mathfrak{D}} \quad \text{on } \partial\Sigma \times (0, \infty),$$
$$u(x, t) \to 0 \quad \text{as } |x| \to \infty.$$

in an exterior domain $\mathfrak{D}^c = \Sigma \subset \mathbb{R}^3$.

We are looking for fundamental solution which will be derived from non-steady case where there is a tensor $\mathcal{Z}(y, z)$ and a three-dimensional vector of distributions $E_{4j}(y, z)$ such that for any vector $a \in \mathbb{R}^3$ the distributions

$$v_z(y) = \mathcal{Z}(y, z)a,$$
$$\pi_z(y) = E_{4j}(y, z)a,$$

solve the system

$$\mathcal{L}v_z + \nabla\pi_z = \delta_z(y)a,$$
$$\text{div } v_z = 0, \qquad (4.2)$$

in the sense of distributions.

The following estimates of $|y - z(t)|$ will play a fundamental role in our argument.

Lemma 4.1 (Key geometrical lemma) *The relation $|e^{-t\cdot\Omega} \cdot v| = |v|$ holds for* $v \in \mathbb{R}^3$.

© Atlantis Press and the author(s) 2016

Š. Nečasová and S. Kračmar, *Navier–Stokes Flow Around a Rotating Obstacle*,
Atlantis Briefs in Differential Equations 3, DOI 10.2991/978-94-6239-231-1_4

Let $R \in (0, \infty)$, $y, z \in B_R$ with $y \neq z$, $t \in (0, \infty)$ with

$$t \leq \min\{|y - z|/(2 \cdot |U|), \ |y - z|/(24 \cdot |\omega| \cdot R), \ (arccos(3/4))/|\omega|\}.$$

Then $|y - z(t)| \geq |y - z|/12$.

Proof The eigenvalues of the matrix Ω are 0, $i \cdot |\omega|$ and $-i \cdot |\omega|$. Thus there is a unitary matrix $A \in \mathbb{C}^{3 \times 3}$ such that $\Omega = A^{-1} \cdot \begin{pmatrix} 0 & 0 & 0 \\ 0 & i \cdot |\omega| & 0 \\ 0 & 0 & -i \cdot |\omega| \end{pmatrix} \cdot A$. Put

$$D(t) := \begin{pmatrix} 1 & 0 & 0 \\ 0 & e^{-i \cdot |\omega| \cdot t} & 0 \\ 0 & 0 & e^{i \cdot |\omega| \cdot t} \end{pmatrix} \quad \text{for } t \in \mathbb{R}.$$

Then $e^{-t \cdot \Omega} = A^{-1} \cdot D(t) \cdot A$. This equation implies that $|e^{-t \cdot \Omega} \cdot v| = |v|$ for $v \in \mathbb{R}^3$.

Abbreviate $\widetilde{y}(t) := A \cdot (y + t \cdot U)$, $\widetilde{z} := A \cdot z$. Note that, in general, the vectors $\widetilde{y}(t)$ and $\widetilde{z}$ are complex-valued. Further observe that

$$|y - z(t)| = |\widetilde{y}(t) - D(t) \cdot \widetilde{z}|, \quad |y + t \cdot U - z| = |\widetilde{y}(t) - \widetilde{z}|.$$

There is an index $j \in \{1, 2, 3\}$ with $|(\widetilde{y}(t) - \widetilde{z})_j| \geq |\widetilde{y}(t) - \widetilde{z}|/3$. If $j = 1$, we get

$$|y - z(t)| = |\widetilde{y}(t) - D(t) \cdot \widetilde{z}| \geq |(\widetilde{y}(t) - D(t) \cdot \widetilde{z})_1| = |(\widetilde{y}(t) - \widetilde{z})_1| \quad (4.3)$$
$$\geq |\widetilde{y}(t) - \widetilde{z}|/3 = |y + t \cdot U - z|/3 \geq (|y - z| - t \cdot |U|)/3 \geq |y - z|/6,$$

where the last inequality holds because $t \leq |y - z|/(2 \cdot |U|)$. Now suppose that $j \in \{2, 3\}$. Without loss of generality, we may take $j = 2$. Then abbreviate

$$a := \widetilde{y}(t)_2, \quad b := \widetilde{z}_2, \quad \gamma := |\omega| \cdot t.$$

We find

$$|y - z(t)|^2 = |\widetilde{y}(t) - D(t) \cdot \widetilde{z}|^2 \geq |(\widetilde{y}(t) - D(t) \cdot \widetilde{z})_2|^2 = |a - e^{i \cdot \gamma} \cdot b|^2 \quad (4.4)$$
$$= (1 - \cos\gamma) \cdot (|a|^2 + |b|^2) + \cos\gamma \cdot |a - b|^2 + 2 \cdot \sin\gamma \cdot (a \bullet (\Im b, -\Re b))$$
$$\geq \cos\gamma \cdot |a - b|^2 - 2 \cdot |\sin\gamma| \cdot |a \bullet (\Im b, -\Re b)|$$
$$= \cos\gamma \cdot |a - b|^2 - 2 \cdot |\sin\gamma| \cdot |(a - b) \bullet (\Im b, -\Re b)|$$
$$\geq \cos\gamma \cdot |a - b|^2 - 2 \cdot |\sin\gamma| \cdot |(a - b)| \cdot |b|,$$

where the symbol $\bullet$ stands for the usual inner product of two vectors in $\mathbb{R}^2$. Next we remark that since $t \leq (arccos(3/4))/|\omega|$, we have $\cos\gamma \geq 3/4$. The assumption $t \leq |y - z|/(24 \cdot |\omega| \cdot R)$ implies

$$|\sin\gamma| \leq \gamma = |\omega| \cdot t \leq |y - z|/(24 \cdot R).$$

(Incidentally, since $t \leq (\arccos(3/4))/|\omega|$, we have $\gamma < \pi/2$, hence $\sin \gamma > 0$, so we might have written $\sin \gamma$ instead of $|\sin \gamma|$.) Moreover $|b| \leq |\widetilde{z}| = |z| \leq R$. Due to these observations, we may conclude from (4.4)

$$|y - z(t)|^2 \geq (3/4) \cdot |a - b|^2 - |y - z| \cdot |a - b|/12. \tag{4.5}$$

On the other hand, recalling that $j = 2$ and using the relation $t \leq |y - z|/(2 \cdot |U|)$, we find

$$|a - b| \geq |\widetilde{y}(t) - \widetilde{z}|/3 = |y + t \cdot U - z|/3 \geq (|y - z| - t \cdot |U|)/3 \geq |y - z|/6,$$

so that $|y - z| \leq 6 \cdot |a - b|$. With this relation, we may conclude from (4.5) that

$$|y - z(t)|^2 \geq (3/4) \cdot |a - b|^2 - |a - b|^2/2 = |a - b|^2/4.$$

Again recalling that $j = 2$ and $t \leq |y - z|/(2 \cdot |U|)$, we finally get

$$|y - z(t)| \geq |a - b|/2 \geq |\widetilde{y}(t) - \widetilde{z}|/6 \geq (|y - z| - t \cdot |U|)/6 \geq |y - z|/12.$$

$\blacksquare$

Theorem 4.1 Let $S, \delta \in (0, \infty)$, $\mu \in (1, \infty)$. Then

$$\int_0^\infty (|y - z(t)|^2 + t)^{-\mu} \, dt \leq \mathfrak{C}(S, \delta, \mu) \, (|y| s_\tau(y))^{-\mu + 1/2} \tag{4.6}$$

for $y \in B_{(1+\delta)S}^c$, $z \in \overline{B_S}$.

Proof Take $y \in B_{(1+\delta)S}^c$, $z \in \overline{B_S}$. We abbreviate $y' := (y_2, y_3)$. In what follows, we will make frequent use of the equation $|e^{-t\Omega} \cdot z| = |z|$; see Lemma 4.1. We will distinguish several cases. To begin with, we suppose that $|y| \leq 8S$. Then, for $t \in (0, \infty)$, we get

$$|y - z(t)|^2 + t = |y - \tau t e_1 - e^{-t\Omega} \cdot z|^2 + t \geq \mathfrak{C}(S)(|y - e^{-t\Omega} \cdot z|^2 + t), \tag{4.7}$$

where we used Lemma 3.7 with $9S$ instead of S. But since $|y| \geq (1+\delta)S$, $|z| \leq S$, we have $|y - e^{-t\Omega} \cdot z| \geq |y| - |z| \geq \delta S$, so that from (4.7),

$$|y - \tau t e_1 - e^{-t\Omega} \cdot z|^2 + t \geq \mathfrak{C}(S, \delta)(1 + t) \quad \text{for } t \in (0, \infty),$$

hence

$$\int_0^\infty (|y - \tau t e_1 - e^{-t\Omega} \cdot z|^2 + t)^{-\mu} \, dt \leq \mathfrak{C}(S, \delta, \mu) \int_0^\infty (1 + t)^{-\mu} \, dt \tag{4.8}$$

$$\leq \mathfrak{C}(S, \delta, \mu) \leq \mathfrak{C}(S, \delta, \mu) |y|^{-2\mu + 1} \leq \mathfrak{C}(S, \delta, \mu) \, (|y| s_\tau(y))^{-\mu + 1/2},$$

with the third inequality following from the assumption $|y| \leq 8\,S$, and the last one from Lemma 3.8. In the rest of this proof, we suppose that $|y| \geq 8\,S$. We note that

$$\int_0^\infty (|y - \tau t e_1 - e^{-t\Omega} \cdot z|^2 + t)^{-\mu}\, dt = \tau^{-1} \int_0^\infty \left(\gamma(y, z, r)^2 + r/\tau\right)^{-\mu}\, dr, \quad (4.9)$$

where we used the abbreviation $\gamma(y, z, r) := |y - r e_1 - e^{-(r/\tau)\Omega} \cdot z|,\ r \in (0, \infty)$. In view of the assumption $|y| \geq 8\,S$, another easy case arises if $y_1 \leq 0$. In fact, we then have

$$\gamma(y, z, r) \geq |y - r e_1| - |z| \geq (|y|^2 + r^2)^{1/2} - S \geq |y|/2 + r/2 - S \geq |y|/4 + r/2$$

for $r \in (0, \infty)$, so that $\gamma(y, z, r)^2 \geq \mathfrak{C}(|y| + r)^2$, hence

$$\int_0^\infty \left(\gamma(y, z, r)^2 + r/\tau\right)^{-\mu}\, dr \leq \mathfrak{C}(\mu) \int_0^\infty (|y| + r)^{-2\mu}\, dr \quad (4.10)$$
$$\leq \mathfrak{C}(\mu) |y|^{-2\mu+1} \leq \mathfrak{C}(S, \mu)\, (|y| s_\tau(y))^{-\mu+1/2},$$

where the last inequality is a consequence of Lemma 3.8. A similar argument holds if $0 \leq y_1 \leq |y|/2$. In fact, since $|y| = (y_1^2 + |y'|^2)^{1/2}$, we then have $|y'| = (|y|^2 - y_1^2)^{1/2} \geq (3|y|^2/4)^{1/2} \geq |y|/2$, so we get for $r \in (0, \infty)$ that

$$\gamma(y, z, t) \geq |y - r e_1|/2 + |y - r e_1|/2 - |z| \geq |y - r e_1|/2 + |y'|/2 - S$$
$$\geq |y - r e_1|/2 + |y|/4 - S \geq |y_1 - r|/2 + |y|/8,$$

where the last inequality follows from the assumption $|y| \geq 8\,S$. We thus get

$$\int_0^\infty \left(\gamma(y, z, r)^2 + r/\tau\right)^{-\mu}\, dr \leq \mathfrak{C}(\mu) \int_0^\infty (|y| + |y_1 - r|)^{-2\mu}\, dr \quad (4.11)$$
$$\leq \mathfrak{C}(\mu) \int_{y_1}^\infty (|y| + r - y_1)^{-2\mu}\, dr \leq \mathfrak{C}(\mu) |y|^{-2\mu+1} \leq \mathfrak{C}(S, \mu)\, (|y| s_\tau(y))^{-\mu+1/2}.$$

The last of the preceding inequalities follows from Lemma 3.8. From now on we suppose that $y_1 \geq |y|/2$. We thus work under the assumption that $y_1 \geq |y|/2 \geq 4\,S$. Then we note

$$\int_0^\infty \left(\gamma(y, z, r)^2 + r/\tau\right)^{-\mu}\, dr \leq \mathfrak{A}_1 + \mathfrak{A}_2, \quad (4.12)$$

with

$$\mathfrak{A}_1 := \int_{y_1 - 2S}^{y_1 + 2S} \left(\gamma(y, z, r)^2 + r/\tau\right)^{-\mu}\, dr,$$

and with $\mathfrak{A}_2$ defined in the same way as $\mathfrak{A}_1$, but with the domain of integration $(y_1 - 2S, y_1 + 2S)$ replaced by $(0, \infty)\backslash(y_1 - 2S, y_1 + 2S)$. We observe that for $r \in (y_1 - 2S, y_1 + 2S)$,

$$r \geq y_1 - 2S \geq |y|/2 - 2S \geq |y|/4,$$

because $y_1 \geq |y|/2$, $|y| \geq 8S$. Therefore

$$\mathfrak{A}_1 \leq \int_{y_1-2S}^{y_1+2S} (r/\tau)^{-\mu}\, dr \leq \mathfrak{C}(\mu)|y|^{-\mu} \int_{y_1-2S}^{y_1+2S} dr \leq \mathfrak{C}(S, \mu)|y|^{-\mu}. \quad (4.13)$$

On the other hand, for $r \in (0, \infty)\backslash(y_1 - 2S, y_1 + 2S)$, we have

$$\gamma(y, z, r) \geq |y - r\,e_1| - |z| \geq |y_1 - r| - S \geq |y_1 - r|/2 + |y_1 - r|/2 - S \geq |y_1 - r|/2,$$

hence

$$\mathfrak{A}_2 \leq \int_{(0,\infty)\backslash(y_1-2S,\, y_1+2S)} \left((|y_1 - r|/2)^2 + r/\tau\right)^{-\mu}\, dr \qquad (4.14)$$

$$\leq \mathfrak{C}(\mu) \int_0^\infty (|y_1 - r| + r^{1/2})^{-2\mu}\, dr$$

$$\leq \mathfrak{C}(\mu) \left(\int_0^{y_1/2} (y_1/2)^{-2\mu}\, dr + \int_{y_1/2}^\infty (|y_1 - r| + (y_1/2)^{1/2})^{-2\mu}\, dr\right)$$

$$\leq \mathfrak{C}(\mu) \left(y_1^{-2\mu+1} + \int_{\mathbb{R}} (|y_1 - r| + y_1^{1/2})^{-2\mu}\, dr\right) \leq \mathfrak{C}(\mu)(y_1^{-2\mu+1} + y_1^{-\mu+1/2})$$

$$\leq \mathfrak{C}(S, \mu)|y|^{-\mu+1/2},$$

with the last inequality following from the assumption $y_1 \geq |y|/2 \geq 4S$. Combining (4.12)–(4.14) yields

$$\int_0^\infty \left(\gamma(y, z, r)^2 + r/\tau\right)^{-\mu}\, dr \leq \mathfrak{C}(S, \mu)|y|^{-\mu+1/2}.$$

Therefore, if $\tau(|y| - y_1) \leq \max\{1, 2\tau S\}$, we have

$$\int_0^\infty \left(\gamma(y, z, r)^2 + r/\tau\right)^{-\mu}\, dr \leq \mathfrak{C}(S, \mu)\,(|y|s_\tau(y))^{-\mu+1/2}. \quad (4.15)$$

Thus we reduced the problem to the case

$$\tau(|y| - y_1) \geq \max\{1, 2\tau S\}, \quad y_1 \geq |y|/2 \geq 4S.$$

Using the relations $\tau(|y| - y_1) \geq 1$, $y_1 \geq 0$, we observe that

$$|y|s_\tau(y) \le |y|2\tau(|y|-y_1) = 2\tau|y||y'|^2/(|y|+y_1) \le 2\tau|y'|^2. \qquad (4.16)$$

We further observe that for $r \in (0,\infty)\setminus(y_1 - 2S, \ y_1 + 2S)$,

$$\gamma(y,z,r) \ge |y - re_1| - |z| \ge |y - re_1|/2 + |y_1 - r|/2 - S \ge |y - re_1|/2$$
$$\ge |y_1 - r|/4 + |y'|/4,$$

so that

$$\mathfrak{A}_2 \le \mathfrak{C}(\mu) \int_{\mathbb{R}} (|y_1 - r| + |y'|)^{-2\mu} \le \mathfrak{C}(\mu)|y'|^{-2\mu+1} \qquad (4.17)$$
$$\le \mathfrak{C}(\mu)\,(|y|s_\tau(y))^{-\mu+1/2},$$

with the last inequality following from (4.16). Using (4.16) again, and recalling that $\tau(|y| - y_1) \ge 2\tau S$, $|y| \ge 4S$, we find for $r \in (0,\infty)$ that

$$\gamma(y,z,r) \ge |y - re_1| - |z| \ge |y'| - S \ge |y'|/2 + \big((2\tau)^{-1}|y|s_\tau(y)\big)^{1/2}/2 - S$$
$$\ge |y'|/2 + (|y|S)^{1/2}/2 - S \ge |y'|/2 + (4S^2)^{1/2}/2 - S = |y'|/2.$$

It follows

$$\mathfrak{A}_1 \le \mathfrak{C}(\mu)|y'|^{-2\mu} \int_{y_1-2S}^{y_1+2S} dr \le \mathfrak{C}(S,\mu)|y'|^{-2\mu} \qquad (4.18)$$
$$\le \mathfrak{C}(S,\mu)\,(|y|s_\tau(y))^{-\mu} \le \mathfrak{C}(S,\mu)\,(|y|s_\tau(y))^{-\mu+1/2},$$

where inequality (4.16) was used once more. By (4.12), (4.17) and (4.18), we see that inequality (4.15) holds also in the case $\tau(|y| - y_1) \ge \max\{1, 2\tau S\}$, $y_1 \ge |y|/2 \ge 4S$. Inequality (4.6) follows with (4.8)–(4.11) and (4.15). As concerns estimate (4.33), it is an immediate consequence of (4.30), Lemma 3.10 and (4.6) with $\mu = -3/2 - |\alpha|/2$. This leaves us to deal with (4.34). In this respect, we remark that the only property of Ω we used in the preceding proof is the relation $|e^{-t\Omega} \cdot x| = |x|$ for $x \in \mathbb{R}^3$, $t \in (0,\infty)$ (Lemma 4.1). Since this relation holds, of course, for any $t \in \mathbb{R}$, and because by Lemma 4.1,

$$|y - t\tau e_1 - e^{-\tau\Omega} \cdot z| = |-z - t\tau e_1 - e^{t\Omega} \cdot (-y)| \quad (y,z \in \mathbb{R}^3, \ t \in \mathbb{R}),$$

we see that we have proved (4.6) also for $z \in B^c_{(1+\delta)S}$ and $y \in \overline{B_S}$, but with y replaced by z on the right-hand side. $\blacksquare$

We present still another estimate of Γ_{jk}:

Lemma 4.2 *Let* $R \in (0,\infty)$, $y \in B_R$, $\epsilon \in (0,\infty)$ *with* $\overline{B_\epsilon(y)} \subset B_R$, $z \in B_R \setminus B_\epsilon(y)$. *Moreover, let* $t \in (0,\infty)$, $j,k \in \{1,2,3\}$, $\alpha \in \mathbb{N}_0^3$ *with* $|\alpha| \le 1$. *Then*

$$|\partial_z^\alpha \Gamma_{jk}(y,z,t)| + |\partial_y^\alpha \Gamma_{jk}(y,z,t)| \le \mathfrak{C}(R,\epsilon) \cdot (\chi_{(0,1]}(t) + \chi_{(1,\infty)}(t) \cdot t^{-3/2}). \qquad (4.19)$$

Proof In order to introduce the fundamental solution constructed by Guenther, Thomann [36] for the linearized variant of (2.1) in a more convenient form for the proof, we define matrices

$$G^{(1)}(y, z, t) := \left(\delta_{jk} - (y - z(t))_j \cdot (y - z(t))_k \cdot |y - z(t)|^{-2} \right)_{1 \leq j, k \leq 3} \cdot e^{-t \cdot \Omega},$$

$$G^{(2)}(y, z, t) := \left(\delta_{jk}/3 - (y - z(t))_j \cdot (y - z(t))_k \cdot |y - z(t)|^{-2} \right)_{1 \leq j, k \leq 3} \cdot e^{-t \cdot \Omega}$$

for $y, z \in \mathbb{R}^3$, $t \in (0, \infty)$ with $y \neq z(t)$.

So, the fundamental solution reads

$$\Gamma(y, z, t) := K\,(y - z(t),\, t) \left[G^{(1)}(y, z, t) - {}_1F_1 \left(1, 5/2, \frac{|y - z(t)|^2}{4t} \right) G^{(2)}(y, z, t) \right]$$

for $y, z \in \mathbb{R}^3$, $t \in (0, \infty)$ with $y \neq z(t)$, $j, k \in \{1, 2, 3\}$.

Let $z \in B_R \backslash B_\epsilon(y)$, $t \in (0, \infty)$. In the case

$$t \leq \min\{\epsilon/(2 \cdot |U|),\ \epsilon/(24 \cdot |\omega| \cdot R),\ (\arccos(3/4))\,/|\omega|,\ \epsilon^2\},$$

we get by referring to Lemma 4.1 that $|y - z(t)| \geq \mathfrak{C} \cdot |y - z| \geq \mathfrak{C} \cdot \epsilon$, hence

$$|y - z(t)|^2/(4 \cdot t) \geq \mathfrak{C} \cdot \epsilon^2/t \geq \mathfrak{C} > 0.$$

Thus by Theorem 3.6,

$$\left| d^k/du^k [e^{-u} \cdot {}_1F_1(1, 5/2, u)] \right|_{u = |y - z(t)|^2/(4 \cdot t)} \leq \mathfrak{C} \cdot (t/\epsilon^2)^{3/2 + k} \quad \text{for } k \in \{0, 1\},$$

and by Lemma 3.1,

$$|\partial_y^\alpha K(y - z(t), t)| \leq \mathfrak{C} \cdot (\epsilon^2 + t)^{-3/2 - |\alpha|/2} \quad \text{for } \alpha \in \mathbb{N}_0^3 \text{ with } |\alpha| \leq 1.$$

Obviously,

$$|\partial_{y_l} G_{jk}^{(i)}(y, z, t)| \leq \mathfrak{C} \cdot |y - z(t)|^{-1} \leq \mathfrak{C} \cdot \epsilon^{-1} \quad \text{for } l \in \{1, 2, 3\}, \ i \in \{1, 2\}.$$

It follows that

$$|\partial_{z_n} \Gamma_{jk}(y, z, t)|$$

$$\leq \mathfrak{C} \cdot \sum_{l=1}^{3} \left(|\partial_{y_l} K(y - z(t), t)| + K(y - z(t), t) \cdot |\partial_{y_l} G_{jk}^{(1)}(y, z, t)| \right.$$

$$+ t^{-3/2} \cdot \left| d/du[e^{-u} \cdot {}_1F_1(1, 5/2, u)] \right|_{u = |y - z(t)|^2/(4 \cdot t)} \cdot |y - z(t)|/t$$

$$+ t^{-3/2} \cdot \left| e^{-u} \cdot {}_1F_1(1, 5/2, u) \right|_{u = |y - z(t)|^2/(4 \cdot t)} \cdot |\partial_{y_l} G_{jk}^{(2)}(y, z, t)|$$

$$\leq \mathfrak{C} \cdot \left((\epsilon^2 + t)^{-2} + (\epsilon^2 + t)^{-3/2} \cdot \epsilon^{-1} + t^{-3/2} \cdot (t/|y - z(t)|^2)^{5/2} \cdot |y - z(t)|/t \right.$$
$$\left. + t^{-3/2} \cdot (t/\epsilon^2)^{3/2} \cdot \epsilon^{-1} \right)$$
$$\leq \mathfrak{C} \cdot \left((\epsilon^2 + t)^{-2} + (\epsilon^2 + t)^{-3/2} \cdot \epsilon^{-1} + \epsilon^{-4} \right) \leq \mathfrak{C}(\epsilon). \qquad (4.20)$$

Next suppose that

$$t \geq \min\{\epsilon/(2 \cdot |U|), \ \epsilon/(24 \cdot |\omega| \cdot R), \ (\arccos(3/4))/|\omega|, \ \epsilon^2\}.$$

Then, if $t \leq 1$, it is clear by Lemma 3.10 that $|\partial_z^\alpha \Gamma_{jk}(y, z, t)| \leq \mathfrak{C}(R, \epsilon)$, and if $t > 1$, it is also clear by the same reference that $|\partial_z^\alpha \Gamma_{jk}(y, z, t)| \leq \mathfrak{C} \cdot t^{-3/2}$. Together we have found that

$$|\partial_z^\alpha \Gamma_{jk}(y, z, t)| \leq \mathfrak{C}(R, \epsilon) \cdot (\chi_{(0,1]}(t) + \chi_{(1,\infty)}(t) \cdot t^{-3/2}).$$

The terms $|\partial_z^\alpha \Gamma_{jk}(y, z, t)|$ and $|\Gamma_{jk}(y, z, t)|$ may be handled in the same way. ∎

Next we give an estimate of the integral $\int_0^\infty \partial_z^\beta \Gamma_{jk}(y, z, t) \, dt$ for $\beta \in \mathbb{N}_0^3$ with $|\beta| \leq 1$. This estimate is interesting mainly in the case when y and z are close to each other.

Theorem 4.2 *Let $R \in (0, \infty)$. Then*

$$\int_0^\infty |\partial_{z_n} \Gamma_{jk}(y, z, t) - \partial_{z_n} \Lambda_{jk}(y - z, t)| \, dt \leq \mathfrak{C}(R) \cdot |y - z|^{-3/2}, \quad (4.21)$$

$$\int_0^\infty |\Gamma_{jk}(y, z, t)| \, dt \leq \mathfrak{C}(R) \cdot |y - z|^{-1},$$

$$\int_0^\infty |\partial_{z_n} \Gamma_{jk}(y, z, t)| \, dt \leq \mathfrak{C}(R) \cdot |y - z|^{-2}$$

for $y, z \in B_R$ with $y \neq z$, $j, k, n \in \{1, 2, 3\}$.

Proof Abbreviate $\epsilon := \min\{C_1 |y - z|, C_2\}$ with C_1, C_2 from Lemma 4.1. Further abbreviate

$$\psi(y, z, t) := e^{t\Omega} \cdot y - \tau t e_1 - z \quad \text{for } t \in (0, \infty), \quad \mathfrak{F}(u) := {}_1F_1(1, 5/2, u) \quad \text{for } u \in \mathbb{R}.$$

Recalling the choice of ϵ, and referring to Lemmas 3.11 and 4.1, we find for $t \in (0, \epsilon)$, $\vartheta \in [0, 1]$ that

$$|\psi(y, z, \vartheta t)| = |y - \tau \vartheta t e_1 - e^{-\vartheta t \Omega} \cdot z| \geq \mathfrak{C} |y - z| \geq \mathfrak{C} \epsilon. \qquad (4.22)$$

Starting from (3.32), we split the left-hand side of (4.21) in the following way:

$$\int_0^\infty |\partial_{z_n}\Gamma_{jk}(y,z,t) - \partial_{z_n}\Lambda_{jk}(y-z,t)|\, dt \tag{4.23}$$

$$\leq \sum_{\nu=1}^9 \int_0^\epsilon \mathfrak{N}_\nu(t)\, dt + \int_\epsilon^\infty |\partial_{z_n}\Gamma_{jk}(y,z,t)|\, dt + \int_\epsilon^\infty |\partial_{z_n}\Lambda_{jk}(y-z,t)|\, dt,$$

with

$$\mathfrak{N}_1(t) := \left| \sum_{l=1}^3 ((e^{-t\Omega})_{jl} - \delta_{jl})\, \partial_{z_n}\Big(\Lambda_{lk}(\psi(y,z,t),\, t)\Big)\right|,$$

$$\mathfrak{N}_2(t) := \left| \partial_{z_n}\Big(K\big(\psi(y,z,t),\, t\big) - K(y-z,t)\Big)\Big(\delta_{jk} - \eta_{jk}(\psi(y,z,t))\Big)\right|,$$

$$\mathfrak{N}_3(t) := \Big| \partial_{z_n}\Big(-K\big(\psi(y,z,t),\, t\big)\mathfrak{F}\big(|\psi(y,z,t)|^2/(4t)\big)$$

$$+ K(y-z,t)\mathfrak{F}\big(|y-z|^2/(4t)\big)\Big)\Big(\delta_{jk}/3 - \eta_{jk}(\psi(y,z,t))\Big)\Big|,$$

$$\mathfrak{N}_4(t) := \left| \partial_{z_n}\big(K(y-z,t)\big)\Big(\eta_{jk}(\psi(y,z,t)) - \eta_{jk}(y-z)\Big)\right|,$$

$$\mathfrak{N}_5(t) := \left| \partial_{z_n}\Big(K(y-z,t)\mathfrak{F}\big(|y-z|^2/(4t)\big)\Big)\Big(\eta_{jk}(\psi(y,z,t)) - \eta_{jk}(y-z)\Big)\right|.$$

The terms $\mathfrak{N}_6(t)$ to $\mathfrak{N}_9(t)$ are defined in the same way as $\mathfrak{N}_2(t)$ to $\mathfrak{N}_5(t)$, respectively, but with the derivative ∂_{z_n} acting on the second factor instead of the first. For example, in the definition of the term $\mathfrak{N}_6(t)$, the derivative is applied to the factor $\delta_{jk} - \eta_{jk}(\psi(y,z,t))$ instead of $K(\psi(y,z,t),\, t) - K(y-z,t)$ as in the definition of $\mathfrak{N}_2(t)$.

In order to estimate $\mathfrak{N}_1(t)$, we observe that the eigenvalues of the matrix Ω are $0,\ i\,|\omega|$ and $-i\,|\omega|$. Therefore there is an invertible matrix $A \in \mathbb{C}^{3\times 3}$ such that

$$\Omega = A \cdot \begin{pmatrix} 0 & 0 & 0 \\ 0 & i\,|\omega| & 0 \\ 0 & 0 & -i\,|\omega| \end{pmatrix} \cdot A^{-1},$$

hence

$$e^{-t\Omega} = A \cdot \begin{pmatrix} 1 & 0 & 0 \\ 0 & e^{-it\,|\omega|} & 0 \\ 0 & 0 & e^{it\,|\omega|} \end{pmatrix} \cdot A^{-1},$$

so for $r,s \in \{1,2,3\}$,

$$|(e^{-t\Omega})_{rs} - \delta_{rs}| \leq \mathfrak{C}\,(|1 - \cos(|\omega|t)| + |\sin(|\omega|t)|) \leq \mathfrak{C}t.$$

Therefore, by Lemma 3.10 and (4.22),

$$\mathfrak{N}_1(t) \le \mathfrak{C}t \left(|\psi(y, z, t)|^2 + t \right)^{-2} \le \mathfrak{C}|\psi(y, z, t)|^{-2} \le \mathfrak{C}\epsilon^{-2},$$

hence $\int_0^\epsilon \mathfrak{N}_1(t) \, dt \le \mathfrak{C}\epsilon^{-1}$. In view of estimating $\mathfrak{N}_2(t)$ to $\mathfrak{N}_9(t)$, we observe that

$$|\partial^\beta \eta_{jk}(x)| \le \mathfrak{C}|x|^{-|\beta|} \quad \text{for } x \in \mathbb{R}^3 \backslash \{0\}, \quad \beta \in \mathbb{N}_0^3 \text{ with } |\beta| \le 2; \tag{4.24}$$

$$|\partial\vartheta \left(|\psi(y, z, \vartheta t)|^2 \right)| = \left| \sum_{m=1}^3 2\psi(y, z, \vartheta t)_m \, t \, (\Omega \cdot e^{\vartheta t \Omega} \cdot y - \tau e_1)_m \right| \tag{4.25}$$

$$\le \mathfrak{C}|\psi(y, z, \vartheta t)|\, t\, (1 + |y|) \le \mathfrak{C}(R)|\psi(y, z, \vartheta t)|\, t \quad \text{for } t \in (0, \epsilon), \ \vartheta \in [0, 1].$$

Similarly,

$$|\partial\vartheta \left(\psi(y, z, \vartheta t)_s \right)| \le \mathfrak{C}(R)t \tag{4.26}$$

for t, ϑ as before and for $s \in \{1, 2, 3\}$. In order to obtain an estimate of $\mathfrak{N}_2(t)$, we apply (4.24), (4.26) and Lemma 3.1 to get

$$\mathfrak{N}_2(t) \le \mathfrak{C} \left| \int_0^1 \sum_{s=1}^3 \partial z_s \partial z_n \left(K \left(\psi(y, z, \vartheta t), \, t \right) \right) \partial\vartheta \left(\psi(y, z, \vartheta t)_s \right) d\vartheta \right|$$

$$\le \mathfrak{C}(R) \int_0^1 \left(|\psi(y, z, \vartheta t)|^2 + t \right)^{-5/2} t \, d\vartheta \quad \text{for } t \in (0, \epsilon).$$

Referring to (4.22), we may conclude that $\mathfrak{N}_2(t) \le \mathfrak{C}(R)(\epsilon^2 + t)^{-3/2}$ for $t \in (0, \epsilon)$, so $\int_0^\epsilon \mathfrak{N}_2(t) \, dt \le \mathfrak{C}(R)\epsilon^{-1}$. Similar arguments yield that $\int_0^\epsilon \mathfrak{N}_6(t) \, dt \le \mathfrak{C}(R)\epsilon^{-3/2}$. Turning to $\mathfrak{N}_3(t)$, we find that

$$\mathfrak{N}_3(t) \tag{4.27}$$

$$\le \mathfrak{C}t^{-3/2} \left| \int_0^1 \partial\vartheta \, \partial z_n \left(e^{-|\psi(y, z, \vartheta t)|^2/(4t)} \, \mathfrak{F} \left(|\psi(y, z, \vartheta t)|^2/(4t) \right) \right) d\vartheta \right|$$

$$= \mathfrak{C}t^{-3/2} \left| \int_0^1 \left([e^{-u} \mathfrak{F}(u)]''_{|u=|\psi(y,z,\vartheta t)|^2/(4t)} \, \psi(y, z, \vartheta t)_n \, (2t)^{-1} \right. \right.$$

$$\partial\vartheta \left(|\psi(y, z, \vartheta t)|^2 \right) (4t)^{-1}$$

$$\left. \left. + [e^{-u} \mathfrak{F}(u)]'_{|u=|\psi(y,z,\vartheta t)|^2/(4t)} \, \partial\vartheta(\psi(y, z, \vartheta t)_n) \, (2t)^{-1} \right) d\vartheta \right|$$

$$\leq \mathfrak{C}(R) t^{-3/2} \int_0^1 \Big(\big| [e^{-u} \mathfrak{F}(u)]'' \big|_{|u=|\psi(y,z,\vartheta t)|^2/(4t)} |\psi(y, z, \vartheta t)|^2 t^{-1}$$

$$+ \big| [e^{-u} \mathfrak{F}(u)]' \big|_{|u=|\psi(y,z,\vartheta t)|^2/(4t)} \Big) \, d\vartheta$$

$$\leq \mathfrak{C}(R) t^{-3/2} \int_0^1 \big(\chi_{(0,1]}(u)(u+1) + \chi_{(1,\infty)}(u) u^{-5/2} \big)_{|u=|\psi(y,z,\vartheta t)|^2/(4t)} \, d\vartheta$$

$$\leq \mathfrak{C}(R) t^{-3/2} \int_0^1 u^{-1}\Big|_{|u=|\psi(y,z,\vartheta t)|^2/(4t)} \, d\vartheta \leq \mathfrak{C}(R) t^{-1/2} \int_0^1 |\psi(y, z, \vartheta t)|^{-2} \, d\vartheta.$$

Note that we applied (4.24) in the first inequality. In the second one, we used (4.25) and (4.26), whereas in the third one, we applied Theorem 3.6. Concerning the next-to-last inequality, we chose the upper bound u^{-1} in order to obtain suitable negative powers of t and $|\psi(y, z, \vartheta t)|$. Making use of (4.22), we may conclude

$$\int_0^\epsilon \mathfrak{N}_3(t) \, dt \leq \mathfrak{C}(R) \epsilon^{-2} \int_0^\epsilon t^{-1/2} \, dt \leq \mathfrak{C}(R) \epsilon^{-3/2}.$$

By exactly the same references and techniques, one may show that

$$\int_0^\epsilon \mathfrak{N}_7(t) \, dt \leq \mathfrak{C}(R) \epsilon^{-3/2}.$$

Next we observe that by (4.22), (4.24) and (4.26),

$$\big| \partial z_n \big(\eta_{jk}(\psi(y, z, t)) - \eta_{jk}(y - z) \big) \big| \tag{4.28}$$

$$= \left| \int_0^1 \sum_{s=1}^3 \partial z_s \partial z_n \big(\eta_{jk}(\psi(y, z, \vartheta t)) \big) \, \partial\vartheta \, (\psi(y, z, \vartheta t)_s) \, d\vartheta \right|$$

$$\leq \mathfrak{C}(R) \int_0^1 |\psi(y, z, \vartheta t)|^{-2} t \, d\vartheta \leq \mathfrak{C}(R) \epsilon^{-2} t.$$

Now we get with Lemma 3.1 that

$$\mathfrak{N}_8(t) \leq \mathfrak{C}(R) (|y - z|^2 + t)^{-3/2} \epsilon^{-2} t \leq \mathfrak{C}(R) \epsilon^{-2} t^{-1/2} \quad \text{for } t \in (0, \epsilon),$$

so that $\int_0^\epsilon \mathfrak{N}_8(t) \, dt \leq \mathfrak{C}(R) \epsilon^{-3/2}$. A similar reasoning yields for $t \in (0, \epsilon)$ that

$$\mathfrak{N}_4(t) \leq \mathfrak{C}(R) (\epsilon^2 + t)^{-2} \epsilon^{-1} t \leq \mathfrak{C}(R) (\epsilon^2 + t)^{-3/2} \epsilon^{-1/2},$$

hence $\int_0^\epsilon \mathfrak{N}_4(t) \, dt \leq \mathfrak{C}(R) \epsilon^{-3/2}$. We find with Theorem 3.6 and (4.28) that

$$\mathfrak{N}_9(t) \leq \mathfrak{C}(R) t^{-3/2} \big| e^{-u} \mathfrak{F}(u) \big|_{|u=|y-z|^2/(4t)} \epsilon^{-2} t$$

$$\leq \mathfrak{C}(R) \epsilon^{-2} t^{-1/2} \big(\chi_{(0,1]}(u) + \chi_{(1,\infty)}(u) u^{-3/2} \big)_{|u=|y-z|^2/(4t)} \leq \mathfrak{C}(R) \epsilon^{-2} t^{-1/2}$$

for $t \in (0, \epsilon)$, hence $\int_0^\epsilon \mathfrak{N}_9(t)\, dt \leq \mathfrak{C}(R)\, \epsilon^{-3/2}$. In the same way we get $\int_0^\epsilon \mathfrak{N}_5(t)\, dt \leq \mathfrak{C}(R)\, \epsilon^{-3/2}$. It is an immediate consequence of Lemma 3.10 that

$$
\int_\epsilon^\infty |\partial_{z_n} \Gamma_{jk}(y, z, t)|\, dt + \int_\epsilon^\infty |\partial_{z_n} \Lambda_{jk}(y - z, t)|\, dt \leq \int_\epsilon^\infty t^{-2}\, dt \leq \mathfrak{C}\epsilon^{-1}.
$$

Thus, in view of (4.23), we have shown that the left-hand side of (4.21) is bounded by $\mathfrak{C}(R)\, \epsilon^{-3/2}$. But since $|y - z| \leq 2R$, and by the choice of ϵ, we have $\epsilon \geq \mathfrak{C}(R)\, |y - z|$, so inequality (4.21) follows.

The other two inequalities stated in the theorem are readily shown. In fact, for $\alpha \in \mathbb{N}_0^3$ with $|\alpha| \leq 1$, we may refer to Lemma 3.10 and (4.22) to obtain

$$
\int_0^\infty |\partial_z^\alpha \Gamma_{jk}(y, z, t)|\, dt \leq \mathfrak{C} \cdot \int_0^\infty (|y - z(t)|^2 + t)^{-3/2 - |\alpha|/2}\, dt
$$

$$
\leq \mathfrak{C} \cdot \left(\int_0^\epsilon (\epsilon^2 + t)^{-3/2 - |\alpha|/2}\, dt + \int_\epsilon^\infty t^{-3/2 - |\alpha|/2}\, dt \right)
$$

$$
\leq \mathfrak{C} \cdot \left(\int_0^\infty (\epsilon^2 + t)^{-3/2 - |\alpha|/2}\, dt + \epsilon^{-1/2 - |\alpha|/2} \right) \leq \mathfrak{C} \cdot \epsilon^{-1 - |\alpha|}.
$$

$\blacksquare$

In view of Theorem 4.2, we may define

$$
\mathcal{Z}_{jk}(y, z) := \int_0^\infty \Gamma_{jk}(y, z, t)\, dt, \quad \mathcal{Y}_{jk}(x) := \int_0^\infty \Lambda_{jk}(x, t)\, dt \quad (4.29)
$$

for $x \in \mathbb{R}^3 \setminus \{0\}$, $y, z \in \mathbb{R}^3$ with $y \neq z$, $j, k \in \{1, 2, 3\}$. The function $(\mathcal{Z}_{jk})_{1 \leq j, k \leq 3}$ is the fundamental solution of (2.2) proposed by Guenther, Thomann in [36].

Remark 5 Theorem 4.2 implies that $\nabla(\mathcal{Z}_{jk} - \mathcal{Y}_{jk})$ is indeed weakly singular with respect to surface integrals in $\mathbb{R}^3$.

Lemma 4.3 *Let $j, k \in \{1, 2, 3\}$. Then $\mathcal{Z}_{jk} \in C^1 \left((\mathbb{R}^3 \times \mathbb{R}^3) \setminus \{(x, x) : x \in \mathbb{R}^3\} \right)$, $\mathcal{Y}_{jk} \in C^1(\mathbb{R}^3 \setminus \{0\})$,*

$$
\partial_{y_n} \mathcal{Z}_{jk}(y, z) = \int_0^\infty \partial_{y_n} \Gamma_{jk}(y, z, t)\, dt, \quad (4.30)
$$

$$
\partial_{z_n} \mathcal{Z}_{jk}(y, z) = \int_0^\infty \partial_{z_n} \Gamma_{jk}(y, z, t)\, dt,
$$

$$
\partial_n \mathcal{Y}_{jk}(x) = \int_0^\infty \partial_{x_n} \Lambda_{jk}(x, t)\, dt
$$

for $y, z \in \mathbb{R}^3$ with $y \neq z$, $x \in \mathbb{R}^3 \setminus \{0\}$, $n \in \{1, 2, 3\}$.

If $R \in (0, \infty)$, $y, z \in B_R$ with $y \neq z$, $\alpha \in \mathbb{N}_0^3$ with $|\alpha| \leq 1$, we have

$$|\partial_y^\alpha \mathcal{Z}_{jk}(y, z)| + |\partial_z^\alpha \mathcal{Z}_{jk}(y, z)| \leq \mathfrak{C}(R) |y - z|^{-1-|\alpha|}. \tag{4.31}$$

Proof Let U, $V \subset \mathbb{R}^3$ be open and bounded, with $\overline{U} \cap \overline{V} \neq \emptyset$. Then $\epsilon :=$ dist$(U, V) > 0$, and there is $R > 0$ with $\overline{U} \cup \overline{V} \subset B_R$. Therefore inequality (4.19) holds for $y \in U$, $z \in V$, $t \in (0, \infty)$. Since $\int_0^\infty \left(\chi_{(0,1)}(t) + \chi_{(1,\infty)}(t) t^{-3/2} \right) dt < \infty$ and in view of Corollary 3.2, the continuous differentiability of $\mathcal{Z}_{jk}$ as well as the first two equations in (4.30) follow by Lebesgue's theorem on dominated convergence. Estimate (4.31) is a consequence of (4.30) and Theorem 4.2. Analogous arguments hold for $\mathcal{Y}_{jk}$. ∎

Lemma 4.4 *Let $j, k \in \{1, 2, 3\}$. For $\alpha, \beta \in \mathbb{N}_0^3$ with $|\alpha + \beta| \leq 2$, $y, z \in \mathbb{R}^3$ with $y \neq z$, the function $(0, \infty) \ni t \mapsto \partial_y^\alpha \partial_z^\beta \Gamma_{jk}(y, z, t) \in \mathbb{R}$ is integrable, the derivative $\partial_y^\alpha \partial_z^\beta \mathcal{Z}_{jk}(y, z)$ exists, and*

$$\partial_y^\alpha \partial_z^\beta \mathcal{Z}_{jk}(y, z) = \int_0^\infty \partial_y^\alpha \partial_z^\beta \Gamma_{jk}(y, z, t) \, dt. \tag{4.32}$$

Moreover, for α, β as before, the derivative $\partial_y^\alpha \partial_z^\beta \mathcal{Z}_{jk}(y, z)$ is a continuous function of $y, z \in \mathbb{R}^3$ with $y \neq z$.

Proof Let $R, \epsilon \in (0, \infty)$ with $\epsilon < R$. Let $C_1 = C_1(R, \tau, \omega)$, $C_2 = C_2(R, \tau, \omega) > 0$ be the constants introduced in Lemma 4.1. Put $\gamma := \min\{C_2, C_1 \cdot \epsilon\}$. For $y, z \in B_R$ with $|y - z| \geq \epsilon$ and for $t \in (0, \gamma]$, we have $t \leq C_2$ and $t \leq C_1 \cdot \epsilon \leq C_1 \cdot |y - z|$, so that $|y - z - \tau \cdot t \cdot e_1 - e^{-t \cdot \Omega} \cdot z| \geq |y - z|/12$ by Lemma 4.1. Thus, referring to Corollary 3.3, we find that for $\alpha, \beta \in \mathbb{N}_0^3$ with $|\alpha + \beta| \leq 2$, $y, z \in B_R$ with $|y - z| \geq \epsilon$,

$$|\partial_y^\alpha \partial_z^\beta \Gamma_{jk}(y, z, t)| \leq \mathfrak{C} \cdot \left(|y - \tau \cdot t \cdot e_1 - e^{-t \cdot \Omega} \cdot z|^2 + t \right)^{-3/2 - |\alpha + \beta|/2}$$

$$\leq \mathfrak{C} \cdot \left(\chi_{(0,\gamma]}(t) \cdot \left((|y - z|/12)^2 + t \right)^{-3/2 - |\alpha + \beta|/2} + \chi_{(\gamma,\infty)}(t) \cdot t^{-3/2 - |\alpha + \beta|/2} \right)$$

$$\leq \mathfrak{C} \cdot \max\{\epsilon^{-3 - |\alpha + \beta|}, \gamma^{-|\alpha + \beta|/2}\} \cdot \left(\chi_{(0,\gamma]}(t) + \chi_{(\gamma,\infty)}(t) \cdot t^{-3/2} \right).$$

But the function $(0, \infty) \ni t \mapsto \chi_{(0,\gamma]}(t) + \chi_{(\gamma,\infty)}(t) \cdot t^{-3/2} \in [0, \infty)$ is integrable. Moreover, for any $t \in (0, \infty)$, we have $\Gamma_{jk}(\cdot, \cdot, t) \in C^2(\mathbb{R}^3 \times \mathbb{R}^3)$ (Corollary 3.2). Therefore Lebesgue's theorem on dominated convergence yields that the statements of Lemma 4.4 hold for $y, z \in \mathbb{R}^3$ with $\epsilon < |y - z| < R$. Since this is true for any $R, \epsilon \in (0, \infty)$ with $\epsilon < R$, Lemma 4.4 is proved. ∎

Now, we are in the position to estimate $\mathcal{Z}_{jk}(y, z)$ in the case that $|z| \leq S$, $|y| \geq (1 + \delta) S$, with $\delta, S > 0$ considered as given quantities. This estimate will play a crucial role in the following.

Theorem 4.3 *Let $S, \delta \in (0, \infty)$, $y \in B^c_{(1+\delta)S}$, $z \in \overline{B_S}$. Then*

$$|\partial^\alpha_y \mathcal{Z}(y, z)| + |\partial^\alpha_z \mathcal{Z}(y, z)| \leq \mathfrak{C}(S, \delta)\, (\,|y|s_\tau(y)\,)^{-1-|\alpha|/2} \qquad (4.33)$$

for $j, k \in \{1, 2, 3\}$, $\alpha \in \mathbb{N}^3_0$ with $|\alpha| \leq 1$. Moreover

$$|\partial^\alpha_y \mathcal{Z}(y, z)| + |\partial^\alpha_z \mathcal{Z}(y, z)| \leq \mathfrak{C}(S, \delta)\, (\,|z|s_\tau(z)\,)^{-1-|\alpha|/2} \qquad (4.34)$$

for $z \in B^c_{(1+\delta)S}$, $y \in \overline{B_S}$, and for j, k, α as in (4.33).

Proof Inequalities (4.33) and (4.34) follow with (4.30), Lemma 3.10, and Theorem 4.1. ∎

Theorem 4.4 *Let $S_1, S \in (0, \infty)$ with $S_1 < S$. Then*

$$|\partial^\alpha_y \partial^\beta_z \mathcal{Z}_{jk}(y, z)| \leq \mathfrak{C}(S_1, S) \cdot (\,|y| \cdot s_\tau(y)\,)^{-1-|\alpha+\beta|/2}$$

for $y \in B^c_S$, $z \in \overline{B_{S_1}}$, $\alpha, \beta \in \mathbb{N}^3_0$ with $|\alpha + \beta| \leq 2$, $1 \leq j, k \leq 3$.

Proof For $y, z, \alpha, \beta, j, k$ as in the theorem, we deduce from (4.32), Corollary 3.3 and Theorem 4.1 that

$$|\partial^\alpha_y \partial^\beta_z \mathcal{Z}_{jk}(y, z)| \leq \int_0^\infty |\partial^\alpha_y \partial^\beta_z \Gamma_{jk}(y, z, t)|\, dt$$

$$\leq \mathfrak{C} \cdot \int_0^\infty \left(|y - \tau \cdot t \cdot e_1 - e^{-t \cdot \Omega} \cdot z|^2 + t\right)^{-3/2 - |\alpha+\beta|/2}\, dt$$

$$\leq \mathfrak{C}(S_1, S) \cdot (\,|y| \cdot s_\tau(y)\,)^{-1-|\alpha+\beta|/2}.$$

∎

Chapter 5
Representation Formula

5.1 Heuristic Approach

The present section is motivated by works of Galdi and Silvestre [33] and [34], where the linear stationary problem (2.2) as well as the nonlinear stationary variant of (2.1),

$$-\Delta u - (U + \omega \times x) \cdot \nabla u + \omega \times u + (u \cdot \nabla)u + \nabla \pi = f, \quad \operatorname{div} u = 0$$
$$\text{in } \mathbb{R}^3 \backslash \overline{\mathfrak{D}} \tag{5.1}$$

are considered. By another transformation of variables, we may suppose there is some $\tau > 0$ with $U = -\tau \cdot (1, 0, 0)$, hence $\omega = \varrho \cdot (1, 0, 0)$ for some $\varrho \in \mathbb{R} \backslash \{0\}$.

It is shown in [33] under suitable assumptions on the data, and in the case of (5.1) additionally under some smallness conditions, that solutions to respectively (2.2) and (5.1) exist in certain Sobolev spaces. These solutions are unique in the space of functions (v, ϱ) satisfying the relation

$$\sup\{|v(x)| \cdot |x| \, : \, x \in \mathbb{R}^3 \backslash B_S\} < \infty \quad \text{for some } S > 0 \text{ with } \overline{\mathfrak{D}} \subset B_S.$$

Article [34] further shows that under additional assumptions on the data, and after some change of variables, the solutions (u, π) constructed in [33] verify the relations

$$\sup\{|u(x)| \cdot |x| \cdot \left(1 + Re \cdot (|x| + x_1)\right) \, : \, x \in \mathbb{R}^3 \backslash B_S\} < \infty, \tag{5.2}$$
$$\sup\{|\nabla u(x)| \cdot |x|^{3/2} \cdot \left(1 + Re \cdot (|x| + x_1)\right)^{3/2} \, : \, x \in \mathbb{R}^3 \backslash B_S\} < \infty.$$

In this section we want to get a more direct access to the decay results in (5.2). In fact, such results are typically deduced from representation formulas.

First we will present the heuristic approach. Let us explain the difficulties in applying the standard approach for deriving representation formulas (or "Green's formulas", as they are also called) for the Eq. (2.2). To this end, consider the Newton potential $\mathcal{N}(x) = (4 \cdot \pi \cdot |x|)^{-1}$, that is, the fundamental solution of the Poisson

© Atlantis Press and the author(s) 2016

Š. Nečasová and S. Kračmar, *Navier–Stokes Flow Around a Rotating Obstacle*,
Atlantis Briefs in Differential Equations 3, DOI 10.2991/978-94-6239-231-1_5

equation. Suppose that $u \in C^2(\mathbb{R}^3 \backslash \mathfrak{D})$. Take $y \in \mathbb{R}^3 \backslash \overline{\mathfrak{D}}$, $R > 0$ with $\overline{\mathfrak{D}} \subset B_R$ and $y \in B_R \backslash \overline{\mathfrak{D}}$. Let $\epsilon > 0$ with $\overline{B_\epsilon(y)} \subset B_R \backslash \overline{\mathfrak{D}}$. Then, following a well known argument, we find

$$\int_{(B_R \backslash \overline{\mathfrak{D}}) \backslash B_\epsilon(y)} (4 \cdot \pi \cdot |y - z|)^{-1} \cdot \Delta u(z) \, dz$$

$$= \int_{\partial B_R \cup \partial \mathfrak{D} \cup \partial B_\epsilon(y)} (4 \cdot \pi)^{-1} \cdot \sum_{k=1}^{3} \left(|y - z|^{-1} \cdot \partial_k u(z) - (y - z)_k \cdot |y - z|^{-3} \cdot u(z) \right)$$
$$\cdot n_k^{(R,\epsilon)}(z) \, do_z,$$

where $n^{(R,\epsilon)}$ denotes the outward unit normal to $(B_R \backslash \overline{\mathfrak{D}}) \backslash B_\epsilon(y)$. By letting ϵ tend to zero, we obtain

$$\int_{B_R \backslash \overline{\mathfrak{D}}} (4 \cdot \pi \cdot |y - z|)^{-1} \cdot \Delta u(z) \, dz$$

$$= \int_{\partial B_R \cup \partial \mathfrak{D}} (4 \cdot \pi)^{-1} \cdot \sum_{k=1}^{3} \left(|y - z|^{-1} \cdot \partial_k u(z) - (y - z)_k \cdot |y - z|^{-3} \cdot u(z) \right)$$
$$\cdot n_k^{(R)}(z) \, do_z \; - \; u(y),$$

with $n^{(R)}$ denoting the outward unit normal to $B_R \backslash \overline{\mathfrak{D}}$. If $u(z)$ decays for $|z| \to \infty$, we may conclude by letting R tend to infinity that

$$\int_{\mathbb{R}^3 \backslash \overline{\mathfrak{D}}} (4 \cdot \pi \cdot |y - z|)^{-1} \cdot \Delta u(z) \, dz$$

$$= \int_{\partial \mathfrak{D}} (4 \cdot \pi)^{-1} \cdot \sum_{k=1}^{3} \left(|y - z|^{-1} \cdot \partial_k u(z) - (y - z)_k \cdot |y - z|^{-3} \cdot u(z) \right) \cdot n_k^{(\mathfrak{D})}(z) \, do_z$$
$$- u(y),$$

hence

$$u(y) = -\int_{\mathbb{R}^3 \backslash \overline{\mathfrak{D}}} (4 \cdot \pi \cdot |y - z|)^{-1} \cdot \Delta u(z) \, dz$$

$$+ \int_{\partial \mathfrak{D}} (4 \cdot \pi)^{-1} \cdot \sum_{k=1}^{3} \left(|y - z|^{-1} \cdot \partial_k u(z) - (y - z)_k \cdot |y - z|^{-3} \cdot u(z) \right) \cdot n_k^{(\mathfrak{D})}(z) \, do_z.$$

Thus the function u is represented as the sum of a volume integral involving Δu, and of a boundary integral involving $u|\partial \mathfrak{D}$ and $\frac{\partial u}{\partial n}|\partial \mathfrak{D}$.

We will proceed in a similar way in order to obtain a representation of u in terms of $-\Delta u - (U + \omega \times z) \cdot \nabla u + \omega \times u + \nabla \pi$, $\operatorname{div} u$, $u \,|\, \partial \mathfrak{D}$, $\frac{\partial u}{\partial n}|\partial \mathfrak{D}$ and

$\pi \mid \partial \mathfrak{D}$, for $u \in C^2(\mathbb{R}^3 \backslash \mathfrak{D})^3$ and $\pi \in C^1(\mathbb{R}^3 \backslash \mathfrak{D})$. In particular, we will not suppose that div $u = 0$. The role of the Newton potential will be played by the fundamental solution of (2.2) mentioned above. But that latter solution is not such a simple function as the Newton potential, so quite an effort is necessary to estimate it in a way which allows us to let ϵ tend to zero (Theorem 5.2) and R to infinity (Theorem 5.3). In particular, the transition of ϵ to zero is difficult to handle.

5.2 Mathematical Preliminaries to the Representation Formula

In this subsection we shall extend properties of $\mathcal{Y}_{jk}$ and $\mathcal{Z}_{jk}$ (for definition see (4.29)) to have estimates which allow us to pass with $\epsilon \to 0$ (see heuristic approach).

The role of the function Λ_{jk} becomes apparent from (4.21) and the ensuing lemma, (see also Theorem 4.2).

Lemma 5.1 *Let* $j, k, n \in \{1, 2, 3\}$, $x \in \mathbb{R}^3 \backslash \{0\}$. *Then*

$$\partial_n \mathcal{Y}_{jk}(x) = \int_0^\infty \partial_{z_n} \Lambda_{jk}(x, t) \, dt$$

$$= -(4 \cdot \pi)^{-1} \cdot |x|^{-3} \cdot \left(\delta_{jk} \cdot x_n + \delta_{jn} \cdot x_k + \delta_{kn} \cdot x_j - 3 \cdot \eta_{jk} \cdot x_n \right)$$

$$+ A \cdot |x|^{-3} \cdot \left(\delta_{jk} \cdot x_n/3 + \delta_{jn} \cdot x_k + \delta_{kn} \cdot x_j - 3 \cdot \eta_{jk} \cdot x_n \right),$$

where the constant A is independent of j, k, n, y *and* z.

This lemma follows with Lemma 3.2. Let us still indicate a simple consequence of the Hölder inequality and Theorem 4.2.

Lemma 5.2 *Let* $j, k \in \{1, 2, 3\}$, $R \in (0, \infty)$, $p \in (3/2, \infty)$, $f \in L^p(B_R)$, $y \in B_R$. *Then*

$$\int_{B_R} |\mathcal{Z}_{jk}(y, z)| |f(z)| \, dz \le \mathfrak{C}(R, p) \cdot \|f\|_p.$$

Proof We know by Theorem 4.2 that

$$\int_{B_R} |\mathcal{Z}_{jk}(y, z)| |f(z)| \, dz \le \int_{B_R} \int_0^\infty |\Gamma_{jk}(y, z, t)| \cdot |f(z)| \, dt \, dz$$

$$\le \mathfrak{C}(R) \cdot \int_{B_R} |y - z|^{-3/2} \cdot |f(z)| \, dz.$$

The lemma follows from this estimate by the Hölder inequality. ∎

A similar argument yields

Lemma 5.3 *Let* $R \in (0, \infty)$, $p \in (3, \infty)$, $f \in L^p(B_R)$, $y \in B_R$. *Then*

$$\int_{B_R} |y - z|^{-2} \cdot |f(y)| \, dy \leq C(R, p) \cdot \|f\|_p.$$

Next, in Lemma 5.4, Theorem 5.1 we prove some technical points. They constituted a major obstacle in the proof of a representation formula for smooth functions $u :$ $\overline{\mathfrak{D}}^c \mapsto \mathbb{R}^3$ in terms of $L(u) + \nabla\pi$, div u and $u|\partial\mathfrak{D}$ (Theorem 6.3). This obstacle consisted in finding a leading term in a decomposition of $\partial z_n \mathcal{Z}_{jk}(y, z)$ such that the remainder term is weakly singular with respect to surface integrals in $\mathbb{R}^3$, see Remark 5. The importance of such a decomposition will become apparent in the proof of Theorem 5.1. The leading term in question is in fact the function $\mathcal{Y}_{jk}(y - z)$, which turns out to coincide with the usual fundamental solution of the Stokes system:

Lemma 5.4 *Let* $j, k \in \{1, 2, 3\}$, $x \in \mathbb{R}^3 \backslash \{0\}$. *Then*

$$\mathcal{Y}_{jk}(x) = (8\pi |x|)^{-1} (\delta_{jk} + x_j x_k |x|^{-2}).$$

Proof Abbreviate $\mathfrak{F}(u) := {}_1F_1(1, 5/2, u)$ for $u \in \mathbb{R}$. Then

$$\mathcal{Y}_{jk}(x) \tag{5.3}$$

$$= \left(\delta_{jk} - \eta_{jk}(x)\right) \int_0^\infty K(x, t) \, dt$$

$$+ \left(-\delta_{jk}/3 + \eta_{jk}(x)\right) (4\pi)^{-3/2} \int_0^\infty t^{-3/2} e^{-|x|^2/(4t)} \mathfrak{F}\left(|x|^2/(4t)\right) dt$$

$$= (4|x|)^{-1} \pi^{-3/2} \left(\left(\delta_{jk} - \eta_{jk}(x)\right) \int_0^\infty s^{-3/2} e^{-1/s} \, ds\right.$$

$$\left. + \left(-\delta_{jk}/3 + \eta_{jk}(x)\right) \int_0^\infty s^{-3/2} e^{-1/s} \mathfrak{F}(1/s) \, ds\right)$$

$$= (4|x|)^{-1} \pi^{-3/2} \left(\left(\delta_{jk} - \eta_{jk}(x)\right) \int_0^\infty t^{-1/2} e^{-t} \, dt\right.$$

$$\left. + \left(-\delta_{jk}/3 + \eta_{jk}(x)\right) \int_0^\infty t^{-1/2} e^{-t} \mathfrak{F}(t) \, dt\right).$$

But $\int_0^\infty t^{-1/2} e^{-t} \, dt = \pi^{1/2}$ by a result about the Gamma function. Therefore, using the abbreviation

$$A := (1/4) \pi^{-3/2} \int_0^\infty t^{-1/2} e^{-t} \mathfrak{F}(t) \, dt,$$

we conclude from (5.3) that

$$\mathcal{Y}_{jk}(x) = (4\pi |x|)^{-1} \left(\delta_{jk} - \eta_{jk}(x) \right) + A |x|^{-1} \left(-\delta_{jk}/3 + \eta_{jk}(x) \right). \quad (5.4)$$

But $\int_0^\infty t^{-1/2} e^{-t} \mathfrak{F}(t)\, dt = 3\pi^{1/2}/2$ as follows by some standard properties of the Gamma function and by the equation $\sum_{n=1}^\infty \left((2n-1)(2n+1) \right)^{-1} = 1/2$. Therefore $A = 3(8\pi)^{-1}$, so the lemma may be deduced from (5.4).   ∎

As a consequence of Theorem 4.2, Lemma 5.1 and the relation $\int_{\partial B_1} \eta_{jk}\, do = 4\pi \delta_{jk}/3$, we obtain

Theorem 5.1 *Let* $j, k, n \in \{1,\, 2,\, 3\}$, $y \in \mathbb{R}^3$, $\epsilon_0 > 0$. *Then, for* $v \in C^0\left(\overline{B_{\epsilon_0}(y)} \right)$,

$$\int_{\partial B_\epsilon(y)} \mathcal{Z}_{jk}(y, z) \cdot (y - z)_n / \epsilon \cdot v(z)\, do_z \to 0 \quad (\epsilon \downarrow 0). \quad (5.5)$$

Let $w \in C^\mu\left(\overline{B_{\epsilon_0}(y)} \right)$ *for some* $\mu \in (0, 1]$. *Then*

$$\int_{B_\epsilon(y)} \sum_{m=1}^3 \partial_{z_m} \mathcal{Z}_{jk}(y, z) \cdot (y - z)_m \cdot \epsilon^{-1} \cdot w(z)\, do_z \to 2 \cdot \delta_{jk} \cdot w(y)/3 \quad (5.6)$$

for $\epsilon \downarrow 0$.

Proof The first relation we get from Lemma 4.3, (4.31). To prove (5.6), we choose $R > 0$ with $\overline{B_{\epsilon_0}(y)} \subset B_R$. For $\epsilon \in (0, \epsilon_0]$, we observe that the difference of the left- and right-hand side of (5.6) is bounded by $\sum_{\nu=1}^3 \mathfrak{N}_\nu(\epsilon)$, with

$$\mathfrak{N}_1(\epsilon) := \int_{\partial B_\epsilon(y)} \sum_{m=1}^3 |\partial_{z_m} \mathcal{Z}_{jk}(y, z)|\, |w(z) - w(y)|\, do_z,$$

$$\mathfrak{N}_2(\epsilon) := |w(y)| \sum_{m=1}^3 \int_{\partial B_\epsilon(y)} |\partial_{z_m} \mathcal{Z}_{jk}(y, z) - \partial_{z_m} \mathcal{Y}_{jk}(y - z)|\, do_z,$$

$$\mathfrak{N}_3(\epsilon) := \left| w(y) \int_{\partial B_\epsilon(y)} \sum_{m=1}^3 \partial_{z_m} \mathcal{Y}_{jk}(y - z)(y - z)_m/\epsilon\, do_z - 2\delta_{jk} w(y)/3 \right|.$$

Put

$$[w]_\mu := \sup\{|w(z) - w(z')| |z - z'|^{-\mu} \; : \; z, z' \in \overline{B_{\epsilon_0}(y)}, \; z \neq z'\}.$$

Let $\epsilon \in (0, \epsilon_0]$. Then with (4.31), we find

$$\mathfrak{N}_1(\epsilon) \le \mathfrak{C}(R)[w]_\mu \int_{\partial B_\epsilon(y)} |y - z|^{-2+\mu}\, do_z \le \mathfrak{C}(R)[w]_\mu \epsilon^\mu.$$

Moreover, referring to (4.30) and to Theorem 4.2, we get

$$\mathfrak{N}_2(\epsilon) \leq \mathfrak{C}(R)\,|w(y)| \int_{\partial B_\epsilon(y)} |y-z|^{-3/2}\, do_z \leq \mathfrak{C}(R)\,|w(y)|\,\epsilon^{1/2}.$$

Using Lemma 5.4, and noting that $\int_{\partial B_1} \eta_{rs}\, do = 4\pi\delta_{rs}/3$ for $r, s \in \{1, 2, 3\}$, we find

$$\int_{\partial B_\epsilon(y)} \sum_{m=1}^{3} \partial_{z_m} \mathcal{Y}_{jk}(y-z)(y-z)_m/\epsilon\, do_z$$

$$= (8\pi)^{-1} \int_{\partial B_1} \sum_{m=1}^{3} \left(\delta_{jk}\eta_{mm} - \delta_{jm}\eta_{km} - \delta_{km}\eta_{jm} + 3\eta_{jk}\eta_{mm} \right) do_\eta$$

$$= 2\delta_{jk}/3,$$

so that $\mathfrak{N}_3(\epsilon) = 0$. Letting ϵ tend to zero, we obtain the theorem. ∎

We remark that the Hölder continuity of w in the preceding theorem is needed in view of Theorem 4.2 so that the relation

$$\int_{B_\epsilon(y)} \sum_{m=1}^{3} \partial_{z_m} \mathcal{Z}_{jk}(y, z) \cdot |w(z) - w(y)|\, do_z \to 0 \quad \text{for } \epsilon \downarrow 0$$

holds.

Also due to the relation $\int_{\partial B_1} \eta_{jk}\, do = (4 \cdot \pi/3) \cdot \delta_{jk}$, we get

Lemma 5.5 *Let $j, k \in \{1, 2, 3\}$, $y \in \mathbb{R}^3$, $\epsilon_0 > 0$, $\alpha \in (0, 1]$, $v \in C^\alpha(\overline{B_{\epsilon_0}(y)})$. Then*

$$\int_{\partial B_\epsilon(y)} (4 \cdot \pi)^{-1} \cdot (y-z)_j \cdot (y-z)_k \cdot \epsilon^{-4} \cdot v(z)\, do_z \to \delta_{jk} \cdot v(y)/3$$

for $\epsilon \downarrow 0$.

5.3 Derivation of the Representation Formula

Now we are in a position to address the representation formula.

First we consider such a formula in a truncated exterior domain $B_R \backslash \overline{\mathfrak{D}}$.

Theorem 5.2 *Let $R \in (0, \infty)$ with $\overline{\mathfrak{D}} \subset B_R$. Let $n^{(R)} : \partial B_R \cup \partial \mathfrak{D} \mapsto \mathbb{R}^3$ denote the outward unit normal to $B_R \backslash \overline{\mathfrak{D}}$. Suppose that $u \in C^2(\overline{B_R \backslash \mathfrak{D}})^3$, $\pi \in C^1(\overline{B_R \backslash \mathfrak{D}})$, and put $f := \mathcal{L}(u) + \nabla \pi$ with $\mathcal{L}$ defined in (2.1). Assume that $y \in B_R \backslash \overline{\mathfrak{D}}$ and $j \in \{1, 2, 3\}$. Then*

$$u_j(y) \tag{5.7}$$

$$= \int_{B_R \setminus \overline{\mathfrak{D}}} \Big(\sum_{k=1}^{3} \mathcal{Z}_{jk}(y, z) \cdot f_k(z)$$

$$+ (4 \cdot \pi)^{-1} \cdot (y - z)_j \cdot |y - z|^{-3} \cdot \operatorname{div} u(z) \Big) \, dz$$

$$+ \int_{\partial B_R \cup \partial \mathfrak{D}} \sum_{k=1}^{3}$$

$$\Big[\sum_{l=1}^{3} \Big(\mathcal{Z}_{jk}(y, z) \cdot \big(\partial_l u_k(z) - \delta_{kl} \cdot \pi(z) + u_k(z) \cdot (U + \omega \times z)_l \big)$$

$$- \partial_{z_l} \mathcal{Z}_{jk}(y, z) \cdot u_k(z) \Big) \cdot n_l^{(R)}(z)$$

$$- (4 \cdot \pi)^{-1} \cdot (y - z)_j \cdot |y - z|^{-3} \cdot u_k(z) \cdot n_k^{(R)}(z) \Big] \, do_z.$$

Proof Let $\epsilon \in (0, \infty)$ with $\overline{B_\epsilon(y)} \subset B_R \setminus \overline{\mathfrak{D}}$. Define $n^{(R,\epsilon)} : \partial B_R \cup \partial \mathfrak{D} \cup \partial B_\epsilon(y) \mapsto \mathbb{R}^3$ by

$$n^{(R,\epsilon)}(z) := n^{(R)}(z) \quad \text{for } z \in \partial B_R \cup \partial \mathfrak{D},$$

$$n^{(R,\epsilon)}(z) := \epsilon^{-1} \cdot (y - z) \quad \text{for } z \in \partial B_\epsilon(y).$$

This means that $n^{(R,\epsilon)}$ is the outward unit normal to $(B_R \setminus \overline{\mathfrak{D}}) \setminus B_\epsilon(y)$. Abbreviate

$$\mathfrak{A}_\epsilon := \int_{(B_R \setminus \overline{\mathfrak{D}}) \setminus B_\epsilon(y)} \sum_{k=1}^{3} \mathcal{Z}_{jk}(y, z) \cdot f_k(z) \, dz.$$

By Lemma 4.2, Lebesgue's theorem and Fubini's theorem, we have

$$\mathfrak{A}_\epsilon = \lim_{S \to \infty, \, \delta \downarrow 0} \int_\delta^S \int_{(B_R \setminus \overline{\mathfrak{D}}) \setminus B_\epsilon(y)} \sum_{k=1}^{3} \Gamma_{jk}(y, z, t) \cdot f_k(z) \, dz \, dt. \tag{5.8}$$

Take δ, $S \in (0, \infty)$ with $\delta < S$ and recall the definition of f in the theorem. Since $\Gamma_{jk}(y, \cdot, t) \in C^\infty(\mathbb{R}^3)$ for $t \in (0, \infty)$ (Corollary 3.2), we may integrate by parts in the integral on the right-hand side of (5.8). Taking into account Theorem 3.4, we thus obtain

$$\mathfrak{A}_\epsilon = \lim_{S \to \infty, \, \delta \downarrow 0} \int_\delta^S \int_{(B_R \setminus \overline{\mathfrak{D}}) \setminus B_\epsilon(y)} \sum_{k=1}^{3} -\partial_t \Gamma_{jk}(y, z, t) \cdot u_k(z) \, dz \, dt + A_{j,\epsilon}(y)$$

with

$$A_{j,\epsilon}(y) := \int_{\partial B_R \cup \partial \mathfrak{D} \cup \partial B_\epsilon(y)} \sum_{k=1}^{3} \sum_{l=1}^{3}$$

$$\left(\mathcal{Z}_{jk}(y,z) \cdot \left(-\partial_l u_k(z) + \delta_{kl} \cdot \pi(z) - (U + \omega \times z)_l \cdot u_k(z) \right) \right.$$

$$\left. + \partial_{z_l} \mathcal{Z}_{jk}(y,z) \cdot u_k(z) \right) \cdot n_l^{(R,\epsilon)}(z) \, do_z.$$

Note that in the integral appearing in the definition of $A_{j,\epsilon}(y)$, we had S tending to infinity and δ to zero and we applied Fubini's theorem. This is possible because by Lemma 4.2, we have

$$\int_0^\infty \int_{\partial B_R \cup \partial \mathfrak{D} \cup \partial B_\epsilon(y)} \left(|\Gamma_{jk}(y,z,t)| + |\partial_{z_n} \Gamma_{jk}(y,z,t)| \right) do_z \, dt < \infty$$

for $1 \leq n \leq 3$. Now it follows that

$$\mathfrak{A}_\epsilon = \lim_{S \to \infty, \, \delta \downarrow 0} \int_{(B_R \setminus \overline{\mathfrak{D}}) \setminus B_\epsilon(y)} \sum_{k=1}^{3} \left(-\Gamma_{jk}(y,z,S) + \Gamma_{jk}(y,z,\delta) \right) \cdot u_k(z) \, dz \quad (5.9)$$

$$+ A_{j,\epsilon}(y).$$

But by Lemma 4.2, Theorem 3.3 and Lebesgue's theorem, we have

$$\int_{(B_R \setminus \overline{\mathfrak{D}}) \setminus B_\epsilon(y)} \sum_{k=1}^{3} \Gamma_{jk}(y,z,S) \cdot u_k(z) \, dz \to 0 \quad \text{for } S \to \infty, \quad (5.10)$$

and

$$\int_{(B_R \setminus \overline{\mathfrak{D}}) \setminus B_\epsilon(y)} \sum_{k=1}^{3} \Gamma_{jk}(y,z,\delta) \cdot u_k(z) \, dz \longrightarrow \quad (5.11)$$

$$\longrightarrow -(4 \cdot \pi)^{-1} \cdot \int_{(B_R \setminus \overline{\mathfrak{D}}) \setminus B_\epsilon(y)} \sum_{k=1}^{3} \left(\delta_{jk} \cdot |y-z|^{-3} - 3 \cdot (y-z)_j \cdot (y-z)_k \cdot |y-z|^{-5} \right)$$

$$\cdot u_k(z) \, dz \quad \text{for } \delta \downarrow 0,$$

Next we again perform an integration by parts, to obtain

$$\int_{(B_R\setminus\overline{\mathfrak{D}})\setminus B_\epsilon(y)} -(4\cdot\pi)^{-1}\cdot\sum_{k=1}^{3}\left(\delta_{jk}\cdot|y-z|^{-3}-3\cdot(y-z)_j\cdot(y-z)_k\right. \quad (5.12)$$

$$\left.\cdot|y-z|^{-5}\right)\cdot u_k(z)\,dz$$

$$=\int_{(B_R\setminus\overline{\mathfrak{D}})\setminus B_\epsilon(y)} -(4\cdot\pi)^{-1}\cdot(y-z)_j\cdot|y-z|^{-3}\cdot\operatorname{div}u(z)\,dz\,+\,B_{j,\epsilon}(y),$$

with

$$B_{j,\epsilon}(y):=\int_{\partial B_R\cup\partial\mathfrak{D}\cup\partial B_\epsilon(y)}(4\cdot\pi)^{-1}\cdot(y-z)_j\cdot|y-z|^{-3}\cdot\sum_{k=1}^{3}u_k(z)\cdot n_k^{(R,\epsilon)}(z)\,do_z.$$

We may conclude from (5.9)–(5.12) that

$$\mathfrak{A}_\epsilon=\int_{(B_R\setminus\overline{\mathfrak{D}})\setminus B_\epsilon(y)} -(4\cdot\pi)^{-1}\cdot(y-z)_j\cdot|y-z|^{-3}\cdot\operatorname{div}u(z)\,dz \quad (5.13)$$

$$+A_{j,\epsilon}(y)+B_{j,\epsilon}(y).$$

Let the terms $\widetilde{A}_j(y)$, $\widetilde{B}_j(y)$ be defined as $A_{j,\epsilon}(y)$ and $B_{j,\epsilon}(y)$, respectively, but with the domain of integration $\partial B_R\cup\partial\mathfrak{D}\cup\partial B_\epsilon$ replaced by $\partial B_R\cup\partial\mathfrak{D}$. Then Theorem 5.1 and Lemma 5.5 yield

$$A_{j,\epsilon}(y)+B_{j,\epsilon}(y) \quad (5.14)$$

$$\longrightarrow \widetilde{A}_j(y)+\widetilde{B}_j(y)+(2/3)\cdot\sum_{k=1}^{3}\delta_{jk}\cdot u_k(y)+(1/3)\cdot\sum_{k=1}^{3}\delta_{jk}\cdot u_k(y)$$

$$=\widetilde{A}_j(y)+\widetilde{B}_j(y)+u_j(y)\quad\text{for }\epsilon\downarrow 0.$$

Since f and ∇u are continuous on $\overline{B_R}\setminus\mathfrak{D}$, Lemmas 5.2 and 5.3 imply

$$\int_{(B_R\setminus\overline{\mathfrak{D}})\setminus B_\epsilon(y)}\sum_{k=1}^{3}\mathcal{Z}_{jk}(y,z)\cdot f_k(y)\,dz \quad (5.15)$$

$$\longrightarrow\int_{B_R\setminus\overline{\mathfrak{D}}}\sum_{k=1}^{3}\mathcal{Z}_{jk}(y,z)\cdot f_k(y)\,dz\quad\text{for }\epsilon\downarrow 0,$$

$$\int_{(B_R\setminus\overline{\mathfrak{D}})\setminus B_\epsilon(y)}(4\cdot\pi)^{-1}\cdot(y-z)_j\cdot|y-z|^{-3}\cdot\operatorname{div}u(z)\,dz \quad (5.16)$$

$$\longrightarrow\int_{B_R\setminus\overline{\mathfrak{D}}}(4\cdot\pi)^{-1}\cdot(y-z)_j\cdot|y-z|^{-3}\cdot\operatorname{div}u(z)\,dz\quad\text{for }\epsilon\downarrow 0.$$

Combining (5.13)–(5.16) yields

$$\int_{B_R \setminus \overline{\mathfrak{D}}} \sum_{k=1}^{3} \mathcal{Z}_{jk}(y, z) \cdot f_k(y) \, dz$$

$$= \int_{B_R \setminus \overline{\mathfrak{D}}} -(4 \cdot \pi)^{-1} \cdot (y - z)_j \cdot |y - {}^t z|^{-3} \cdot \operatorname{div} u(z) \, dz + \widetilde{A}_j(y) + \widetilde{B}_j(y) + u_j(y).$$

This proves (5.7).   ∎

By letting R tend to infinity in (5.7), the following result may be deduced from Theorem 5.2:

Theorem 5.3 *Let $u \in C^2(\mathbb{R}^3 \setminus \mathfrak{D})^3$, $\pi \in C^1(\mathbb{R}^3 \setminus \mathfrak{D})$, $f \in C^0(\mathbb{R}^3 \setminus \mathfrak{D})^3$ with $f = L(u) + \nabla \pi$. Suppose there is $S > 0$ with $\overline{\mathfrak{D}} \subset B_S$ such that*

$$\int_{\mathbb{R}^3 \setminus B_S} |z|^{-1/2} \cdot |f(z)| \, dz < \infty, \qquad \int_{\mathbb{R}^3 \setminus B_S} |z|^{-2} \cdot |\operatorname{div} u(z)| \, dz < \infty.$$

Further suppose there is a sequence (R_n) in (S, ∞) such that

$$R_n^{-1/2} \cdot \int_{\partial B_{R_n}} \left(|\nabla u(z)| + |\pi(z)| + |u(z)| \right) do_z + R_n^{-2} \cdot \int_{\partial B_{R_n}} |\operatorname{div} u(z)| \, do_z \longrightarrow 0$$

for $n \to \infty$. Let $j \in \{1, 2, 3\}$, $y \in \mathbb{R}^3 \setminus \overline{\mathfrak{D}}$. Then

$$u_j(y)$$

$$= \int_{\mathbb{R}^3 \setminus \overline{\mathfrak{D}}} \Big(\sum_{k=1}^{3} \mathcal{Z}_{jk}(y, z) \cdot f_k(z)$$

$$+ (4 \cdot \pi)^{-1} \cdot (y - z)_j \cdot |y - z|^{-3} \cdot \operatorname{div} u(z) \Big) \, dz$$

$$- \int_{\partial \mathfrak{D}} \sum_{k=1}^{3}$$

$$\left[\sum_{l=1}^{3} \Big(\mathcal{Z}_{jk}(y, z) \cdot \big(\partial_l u_k(z) - \delta_{kl} \cdot \pi(z) + u_k(z) \cdot (U + \omega \times z)_l \big) \right.$$

$$- \partial_{z_l} \mathcal{Z}_{jk}(y, z) \cdot u_k(z) \Big) \cdot n_l^{(\mathfrak{D})}(z)$$

$$\left. - (4 \cdot \pi)^{-1} \cdot (y - z)_j \cdot |y - z|^{-3} \cdot u_k(z) \cdot n_k^{(\mathfrak{D})}(z) \right] do_z.$$

Chapter 6
Asymptotic Behavior

6.1 Some Volume Potentials

In the present section, we study the volume potentials which arise. There are two types of such potentials, involving the kernels $\mathcal{Z}_{jk}$ and E_{4j}, respectively. We begin by considering the potential related to $\mathcal{Z}_{jk}$.

Lemma 6.1 *Let $p \in (1, \infty)$, $q \in (1, 2)$, $f \in L^p_{\text{loc}}(\mathbb{R}^3)^3$ with $f|B^c_S \in L^q(B^c_S)^3$ for some $S \in (0, \infty)$. Then, for $j, k \in \{1, 2, 3\}$, $\alpha \in \mathbb{N}^3_0$ with $|\alpha| \leq 1$, we have*

$$\int_{\mathbb{R}^3} |\partial^\alpha_y \mathcal{Z}_{jk}(y, z)| \, |f_k(z)| \, dy < \infty \quad \text{for a.e. } y \in \mathbb{R}^3. \tag{6.1}$$

We define $\mathfrak{R}(f) : \mathbb{R}^3 \mapsto \mathbb{R}^3$ by

$$\mathfrak{R}_j(f)(y) := \int_{\mathbb{R}^3} \sum_{k=1}^3 \mathcal{Z}_{jk}(y, z) \, f_k(z) \, dz$$

for $y \in \mathbb{R}^3$ such that (6.1) holds; else we set $\mathfrak{R}_j(f)(y) := 0$ ($1 \leq j \leq 3$). Then $\mathfrak{R}(f) \in W^{1,1}_{\text{loc}}(\mathbb{R}^3)^3$ and

$$\partial_l \mathfrak{R}_j(f)(y) := \int_{\mathbb{R}^3} \sum_{k=1}^3 \partial_{y_l} \mathcal{Z}_{jk}(y, z) \, f_k(z) \, dz \tag{6.2}$$

for $j, l \in \{1, 2, 3\}$ and for a. e. $y \in \mathbb{R}^3$. Moreover, for $R \in (0, \infty)$ we have

$$\|\mathfrak{R}(f|B_R) \, | \, B_R\|_p \leq \mathfrak{C}(R, p) \, \|f|B_R\|_p. \tag{6.3}$$

© Atlantis Press and the author(s) 2016
Š. Nečasová and S. Kračmar, *Navier–Stokes Flow Around a Rotating Obstacle*,
Atlantis Briefs in Differential Equations 3, DOI 10.2991/978-94-6239-231-1_6

Proof Take j, k, α as in (6.1). Let $R \in (0, \infty)$. Then we find by (4.31) that

$$\int_{B_R} |\partial_y^\alpha \mathcal{Z}_{jk}(y, z)| \, dz \leq \mathfrak{C}(R) \int_{B_R} |y - z|^{-1-|\alpha|} \, dz \leq \mathfrak{C}(R) \int_{B_{2R}(y)} |y - z|^{-1-|\alpha|} \, dz$$
$$\leq \mathfrak{C}(R)$$

for $y \in B_R$, and analogously $\int_{B_R} |\partial_y^\alpha \mathcal{Z}_{jk}(y, z)| \, dy \leq \mathfrak{C}(R)$ for $z \in B_R$. It follows by Hölder's inequality that

$$\left(\int_{B_R} \left(\int_{B_R} |\partial_y^\alpha \mathcal{Z}_{jk}(y, z)| \, |f_k(z)| \, dz \right)^p dy \right)^{1/p} \tag{6.4}$$
$$\leq \left(\int_{B_R} \left(\int_{B_R} |\partial_y^\alpha \mathcal{Z}_{jk}(y, z)| \, dz \right)^{p-1} \left(\int_{B_R} |\partial_y^\alpha \mathcal{Z}_{jk}(y, z)| \, |f(z)|^p \, dz \right) dy \right)^{1/p}$$
$$\leq \mathfrak{C}(R, p) \left(\int_{B_R} \int_{B_R} |\partial_y^\alpha \mathcal{Z}_{jk}(y, z)| \, |f(z)|^p \, dz \, dy \right)^{1/p} \leq \mathfrak{C}(R, p) \| f | B_R \|_p.$$

This means in particular that the integral $\int_{B_n} |\partial_y^\alpha \mathcal{Z}_{jk}(y, z)| \, |f_k(z)| \, dz$ is finite for a.e. $y \in B_n$, $n \in \mathbb{N}$, and that inequality (6.3) is proved.

Once again take j, k, α as in (6.1), and let $n \in \mathbb{N}$ with $n \geq S$. Then, using (4.34) with S replaced by $n/2$ and with $\delta = 1/2$, we find for $y \in B_{n/2}$ that

$$\int_{B_n^c} |\partial_y^\alpha \mathcal{Z}_{jk}(y, z)| \, |f_k(z)| \, dz \leq \mathfrak{C}(n) \int_{B_n^c} \left(|z| s_\tau(z) \right)^{-1-|\alpha|/2} |f(z)| \, dz \tag{6.5}$$
$$\leq \mathfrak{C}(n) \left(\int_{B_n^c} \left(|z| s_\tau(z) \right)^{-q'} dz \right)^{1/q'} \| f | B_n^c \|_q \leq \mathfrak{C}(n, q) \| f | B_S^c \|_q,$$

where the last inequality holds due to Theorem 3.5 and the assumption $q < 2$ (hence $q' > 2$). We thus have shown that the relation in (6.1) holds for a. e. $y \in B_{n/2}$. Since this is true for any $n \in \mathbb{N}$ with $n \geq S$, (6.1) is proved. We deduce from (6.4) and (6.5) that

$$\int_{B_{n/2}} \int_{\mathbb{R}^3} \left| \sum_{k=1}^3 \partial_y^\alpha \mathcal{Z}_{jk}(y, z) f_k(z) \right| dz \, dy \leq \mathfrak{C}(n, p, q) \left(\| f | B_n \|_p + \| f | B_S^c \|_q \right) \tag{6.6}$$

for $n \in \mathbb{N}$ with $n \geq S$. This means that $\mathfrak{R}_j(f) \in L_{1,\text{loc}}(\mathbb{R}^3)$, and that the function associating a. e. $y \in \mathbb{R}^3$ with the integral $\int_{\mathbb{R}^3} \sum_{k=1}^3 \partial_{y_l} \mathcal{Z}_{jk}(y, z) f_k(z) \, dz$ also belongs to $L_{1,\text{loc}}(\mathbb{R}^3)$ for $1 \leq l \leq 3$. Now take $\Phi \in C_0^\infty(\mathbb{R}^3)^3$. Then, by (6.6) and because the support of Φ is compact,

$$\int_{\mathbb{R}^3} \partial_l \Phi(y) \mathfrak{R}_j(f)(y) \, dy = \sum_{k=1}^3 \lim_{\epsilon \downarrow 0} \int_{\mathbb{R}^3} \int_{\mathbb{R}^3 \setminus B_\epsilon(z)} \partial_l \Phi(y) \mathcal{Z}_{jk}(y, z) \, dy \, f_k(z) \, dz.$$

$$\tag{6.7}$$

But for any $\epsilon > 0$, we may perform a partial integration in the inner integral on the right-hand side of (6.7) (first statement of Lemma 4.3). Due to (4.31), the term with a surface integral on $\partial B_\epsilon(z)$ arising in this way tends to zero for $\epsilon \downarrow 0$. (Note that for $\epsilon \in (0, 1]$, say, and for $y \in \partial B_\epsilon(z)$, the integral with respect to z only extends over B_{n+1}, if $n \in \mathbb{N}$ is chosen so large that $\mathrm{supp}(\Phi) \subset B_n$.) After letting ϵ tend to zero, we obtain an equation which implies that $\mathfrak{R}_j(f) \in W_{\mathrm{loc}}^{1,1}(\mathbb{R}^3)$ and equation (6.2) holds. ∎

Lemma 6.2 *Take* p, q, f *as in Lemma 6.1, and suppose in addition that* $p > 3/2$. *Then the relation in (6.1) holds for any* $y \in \mathbb{R}^3$ *(without the restriction "a. e."), and the function* $\mathfrak{R}(f)$ *is continuous.*

Proof We show that $\mathfrak{R}(f)$ is continuous. The relation in (6.1) for any $y \in \mathbb{R}^3$ may be established by a similar but simpler argument.

Let $j \in \{1, 2, 3\}$, $R \in (S, \infty)$. It suffices to prove that that $\mathfrak{R}_j(f) | B_R$ is continuous. But for $z \in B_{2R}^c$, $y \in B_R$, we get by (4.34) that

$$\left| \sum_{k=1}^{3} \mathcal{Z}_{jk}(y, z) f_k(z) \right| \leq \mathfrak{C}(R) \left(|z| s_\tau(z) \right)^{-1} |f(z)|.$$

Since by a computation as in (6.5), the function

$$\mathbb{R}^3 \ni z \mapsto \chi_{B_{2R}^c}(z) \left(|z| s_\tau(z) \right)^{-1} |f(z)| \in [0, \infty)$$

is integrable, we may conclude in view of the first statement of Lemma 4.3 that the integral $\int_{B_{2R}^c} \sum_{k=1}^{3} \mathcal{Z}_{jk}(y, z) f_k(z) \, dz$ is continuous as a function of $y \in B_R$. Thus we still have to show that the function

$$I(y) := \int_{B_{2R}} \sum_{k=1}^{3} \mathcal{Z}_{jk}(y, z) f_k(z) \, dz \quad (y \in B_R)$$

is continuous as well. So take y, $y' \in B_R$ with $y \neq y'$. Then

$$|I(y) - I(y')| \leq \mathfrak{N}_1 + \mathfrak{N}_2, \tag{6.8}$$

with

$$\mathfrak{N}_1 := \sum_{x \in \{y, y'\}} \int_{B_R \cap A} \sum_{k=1}^{3} |\mathcal{Z}_{jk}(x, z) f_k(z)| \, dz,$$

$$\mathfrak{N}_2 := \int_{B_R \setminus A} \left| \int_0^1 \sum_{k,l=1}^{3} \partial_{x_l} \mathcal{Z}_{jk}(x, z)_{|x=y'+\vartheta(y-y')} (y - y')_l \, d\vartheta \right| |f_k(z)| \, dz,$$

with $A := B_{2|y-y'|}(y)$. We get with (4.31) that

$$\mathfrak{N}_1 \le \mathfrak{C}(R) \sum_{x \in \{y,y'\}} \int_{B_R \cap A} |x - z|^{-1} |f(z)| \, dz$$

$$\le \mathfrak{C}(R) \sum_{x \in \{y,y'\}} \left(\int_{B_{3|y-y'|}(x)} |x - z|^{-p'} \, dz \right)^{1/p'} \|f|B_R\|_p.$$

Since $p > 3/2$, hence $p' < 3$ and we may conclude that $\mathfrak{N}_1 \le \mathfrak{C}(R)|y - y'|^{-1+3/p'} \|f|B_R\|_p$, with $-1 + 3/p' > 0$. In order to estimate $\mathfrak{N}_2$, we note that

$$|y' + \vartheta(y - y') - z| \ge |y - z| - |y - y'| \ge |y - z|/2 \ge |y - y'|$$

for $z \in \mathbb{R}^3 \backslash A$, $\vartheta \in [0, 1]$. Therefore by (4.31), if $2p' > 3$,

$$\mathfrak{N}_2 \le \mathfrak{C}(R)|y - y'| \left(\int_{B_R \backslash A} |y - z|^{-2p'} \, dz \right)^{1/p'} \|f|B_R\|_p$$

$$\le \mathfrak{C}(R)|y - y'|^{-1+3/p'} \|f|B_R\|_p.$$

In the case $2p' < 3$, the factor $|y - y'|^{-1+3/p'}$ on the right-hand side of the preceding inequality may be replaced by $|y - y'|$ and in the case $2p' = 3$ by $|y - y'| \ln(|y - y'|/(2R))$. In view of (6.8), we have thus shown that $I(y)$ is a continuous function of $y \in B_R$. This completes the proof of Lemma 6.2. $\blacksquare$

6.2 Asymptotic Profile

The crucial idea of the proof of the next theorem consists in reducing an estimate of $\mathfrak{R}(f)$ to an estimate of a convolution integral involving an upper bound of an Oseen fundamental solution.

Theorem 6.1 *Let* S, S_1, $\gamma \in (0, \infty)$ *with* $S_1 < S$, $p \in (1, \infty)$, $A \in [2, \infty)$, $B \in \mathbb{R}$, $f : \mathbb{R}^3 \mapsto \mathbb{R}^3$ *measurable with*

$$f|B_{S_1} \in L^p(B_{S_1})^3, \quad |f(z)| \le \gamma |z|^{-A} s_\tau(z)^{-B} \ \text{for} \ z \in B_{S_1}^c, \quad A + \min\{1, B\} \ge 3.$$

Let $i, j \in \{1, 2, 3\}$, $y \in B_S^c$. *Then*

$$|\mathfrak{R}_j(f)(y)| \le \mathfrak{C}(S, S_1, A, B) (\|f|B_{S_1}\|_1 + \gamma) \left(|y| s_\tau(y) \right)^{-1} l_{A,B}(y), \quad (6.9)$$

$$|\partial_{y_i} \mathfrak{R}_j(f)(y)| \le \mathfrak{C}(S, S_1, A, B) (\|f|B_{S_1}\|_1 + \gamma) \quad (6.10)$$
$$\left(|y| s_\tau(y) \right)^{-3/2} s_\tau(y)^{\max(0, \, 7/2 - A - B)} l_{A,B}(y),$$

where $l_{A,B}(y) = \begin{cases} 1 & \text{if } A + \min\{1, B\} > 3, \\ \max(1, \ln|y|) & \text{if } A + \min\{1, B\} = 3. \end{cases}$

Proof By (4.33) with S, δ replaced by S_1, $S/S_1 - 1$, respectively, we find for $k \in \{1, 2, 3\}$, $\alpha \in \mathbb{N}_0^3$ with $|\alpha| \leq 1$, that

$$\int_{B_{S_1}} |\partial_y^\alpha \mathcal{Z}_{jk}(y, z)| \, |f(z)| \, dz \leq \mathfrak{C}(S, S_1)\left(|y| s_\tau(y)\right)^{-1-|\alpha|/2} \|f|B_{S_1}\|_1. \quad (6.11)$$

Recalling Lemmas 4.3, 3.10, 3.11 and 4.1 we see that

$$\mathfrak{A}_\alpha := \int_{B_{S_1}^c} |\partial_y^\alpha \mathcal{Z}_{jk}(y, z)| \, |f(z)| \, dz$$

$$\leq \mathfrak{C}\gamma \int_0^\infty \int_{B_{S_1}^c} (|y - \tau t e_1 - e^{-t\Omega} \cdot z|^2 + t)^{-3/2-|\alpha|/2} |z|^{-A} s_\tau(z)^{-B} \, dz \, dt$$

$$= \mathfrak{C}\gamma \int_0^\infty \int_{B_{S_1}^c} (|y - \tau t e_1 - x|^2 + t)^{-3/2-|\alpha|/2} |x|^{-A} s_\tau(e^{t\Omega} \cdot x)^{-B} \, dx \, dt$$

$$= \mathfrak{C}\gamma \int_{B_{S_1}^c} \int_0^\infty (|y - \tau t e_1 - x|^2 + t)^{-3/2-|\alpha|/2} \, dt \, |x|^{-A} s_\tau(x)^{-B} \, dx,$$

where the last equation holds due Lemmas 3.11 and 4.1. Now we apply (4.6) with y replaced by $y - x$ and with $z = 0$. Moreover we use Theorem 4.2. It follows

$$\mathfrak{A}_\alpha \leq \mathfrak{C}(S)\gamma\left(\int_{B_{S_1}^c \cap B_{S/2}(y)} |y - x|^{-1-|\alpha|} |x|^{-A} s_\tau(x)^{-B} \, dx \right. \quad (6.12)$$

$$\left. + \int_{B_{S_1}^c \setminus B_{S/2}(y)} \left(|y - x| s_\tau(y - x)\right)^{-1-|\alpha|/2} |x|^{-A} s_\tau(x)^{-B} \, dx\right).$$

Next we observe that for $x \in B_{S/2}(y)$, we have $|x| \geq |y| - |y - x| \geq |y| - S/2 \geq |y|/2$,

$$s_\tau(x)^{-1} \leq \mathfrak{C}(1 + |y - x|) s_\tau(y)^{-1} \leq \mathfrak{C}(S) s_\tau(y)^{-1}$$

(see Lemma 3.6), and similarly $s_\tau(y)^{-1} \leq \mathfrak{C}(S) s_\tau(x)^{-1}$. For $x \in B_{S/2}(y)^c$, we find

$$|y - x| = |y - x|/2 + |y - x|/2 \geq S/4 + |y - x|/2 \geq \min\{S/4, 1/2\}(1 + |y - x|).$$

Thus, independently of the sign of B, we may conclude from (6.12) that

$$\mathfrak{A}_\alpha \le \mathfrak{C}(S, S_1, A, B)\gamma\left(|y|^{-A} s_\tau(y)^{-B} \int_{B_{S/2}(y)} |y - x|^{-1-|\alpha|}\, dx\right. \tag{6.13}$$

$$+ \left.\int_{B_{S_1}^c \setminus B_{S/2}(y)} \left((1 + |y - x|) s_\tau(y - x)\right)^{-1-|\alpha|/2} (1 + |x|)^{-A} s_\tau(x)^{-B}\, dx\right)$$

$$\le \mathfrak{C}(S, S_1, A, B)\gamma\left(|y|^{-A} s_\tau(y)^{-B}\right.$$

$$+ \left.\int_{\mathbb{R}^3} \left((1 + |y - x|) s_\tau(y - x)\right)^{-1-|\alpha|/2} (1 + |x|)^{-A} s_\tau(x)^{-B}\, dx\right).$$

In the case $\alpha = 0$, we refer to the proof of [41, Theorem 3.1] and our assumptions on A and B to deduce from (6.13) that

$$\mathfrak{A}_0 \le \mathfrak{C}(S, S_1, A, B)\gamma\left(|y|^{-A} s_\tau(y)^{-B} + \left(|y| s_\tau(y)\right)^{-1} l_{A,B}(y)\right). \tag{6.14}$$

But by Lemma 3.8 and because $A - 3/2 > 0$, $A + B \ge A + \min\{1, B\} \ge 3$, we have

$$|y|^{-A} s_\tau(y)^{-B} \le \mathfrak{C}(S, A)|y|^{-3/2} s_\tau(y)^{-A+3/2-B} \tag{6.15}$$

$$\le \mathfrak{C}(S, A)|y|^{-3/2} s_\tau(y)^{-3/2},$$

so we may conclude from (6.14) that

$$\mathfrak{A}_0 \le \mathfrak{C}(S, S_1, A, B)\gamma\left(|y| s_\tau(y)\right)^{-1} l_{A,B}(y).$$

Inequality (6.9) follows from (6.11) and the preceding estimate. If $|\alpha| = 1$, Eq. (6.13) and the proof of [41, Theorem 3.2] yield

$$\mathfrak{A}_\alpha \le \mathfrak{C}(S, S_1, A, B)\gamma \tag{6.16}$$

$$\left(|y|^{-A} s_\tau(y)^{-B} + \left(|y| s_\tau(y)\right)^{-3/2} s_\tau(y)^{\max(0,\, 7/2-A-B)} l_{A,B}(y)\right).$$

Hence by (6.15),

$$\mathfrak{A}_\alpha \le \mathfrak{C}(S, S_1, A, B)\gamma\left(|y| s_\tau(y)\right)^{-3/2} s_\tau(y)^{\max(0,\, 7/2-A-B)} l_{A,B}(y).$$

This estimate together with (6.11) implies (6.10). ∎

Now we turn to volume integrals involving the kernel E_{4j}.

Lemma 6.3 *Let $p \in (1, \infty)$, $q \in (1, 3)$, $g \in L_{loc}^p(\mathbb{R}^3)$ with $g|B_S^c \in L^q(B_S^c)$ for some $S \in (0, \infty)$. Then, for $j \in \{1, 2, 3\}$,*

$$\int_{\mathbb{R}^3} |E_{4j}(y - z)|\, |g(z)|\, dy < \infty \quad \text{for a.e. } y \in \mathbb{R}^3. \tag{6.17}$$

Thus we may define $\mathfrak{S}(g) : \mathbb{R}^3 \mapsto \mathbb{R}^3$ *by*

$$\mathfrak{S}_j(g)(y) := \int_{\mathbb{R}^3} E_{4j}(y - z) g(z) \, dz$$

for $y \in \mathbb{R}^3$ *such that (6.17) holds, and* $\mathfrak{S}_j(g)(y) := 0$ *else, where*$(1 \le j \le 3)$*. Then* $\mathfrak{S}(g) \in W_{loc}^{1,1}(\mathbb{R}^3)^3$*. For* $R \in (0, \infty)$ *we have*

$$\|\mathfrak{S}(g|B_R) \mid B_R\|_p \le \mathfrak{C}(R, p) \|g|B_R\|_p. \tag{6.18}$$

If $p > 3$, *the relation in (6.17) holds for any* $y \in \mathbb{R}^3$ *(without the restriction "a.e."), and* $\mathfrak{S}(g)$ *is continuous.*

Proof Lemma 6.3 may be shown by arguments analogous to those we used to prove Lemmas 6.1 and 6.2, except as concerns the claim $\mathfrak{S}(g) \in W_{loc}^{1,1}(\mathbb{R}^3)^3$. To establish this latter point, a different reasoning based on the Calderon–Zygmund inequality is needed because the derivative $\partial_l E_{4j}$ is a singular kernel in $\mathbb{R}^3$. We refer to [27, Sect. 4.2] for details. ∎

Theorem 6.2 *Let* $S, S_1, \widetilde{\gamma} \in (0, \infty)$ *with* $S_1 < S$, $p \in (1, \infty)$, $C \in (5/2, \infty)$, $D \in \mathbb{R}$, $g : \mathbb{R}^3 \mapsto \mathbb{R}$ *measurable with*

$$g|B_{S_1} \in L^p(B_{S_1}), \quad |g(z)| \le \widetilde{\gamma} |z|^{-C} s_\tau(z)^{-D} \quad \text{for } z \in B_{S_1}^c, \quad C + \min\{1, D\} > 3.$$

Let $j \in \{1, 2, 3\}$, $y \in B_S^c$. *Then*

$$|\mathfrak{S}_j(g)(y)| \le \mathfrak{C}(S, S_1, C, D) (\|g|B_{S_1}\|_1 + \widetilde{\gamma}) |y|^{-2}. \tag{6.19}$$

If supp$(g) \subset B_{S_1}$, *we further have*

$$|\partial_n \mathfrak{S}_j(g)(y)| \le \mathfrak{C}(S, S_1) \|g\|_1 |y|^{-3} \quad (1 \le n \le 3). \tag{6.20}$$

Proof Inequality (6.19) may be proved in the same way as Theorem 6.1, except that the reference to [41, Theorems 3.1, 3.2] is replaced by [41, Theorem 3.4] and that the argument becomes simpler due to the much simpler structure of the kernel E_{4j} compared to $\mathcal{Z}_{jk}$. As concerns (6.20), observe that $|y - z| \ge (1 - S_1/S)|y|$ for $z \in B_{S_1}$, so if supp$(g) \subset B_{S_1}$, it is obvious that

$$\mathfrak{S}_j(g)|B_S^c \in C^1(B_S^c), \quad \int_{B_{S_1}} |\partial_l E_{4j}(y - z)| |g(z)| \, dz < \infty,$$

$$\partial_l \mathfrak{S}_j(g)(y) = \int_{B_{S_1}} \partial_l E_{4j}(y - z) g(z) \, dz \quad (1 \le l \le 3).$$

Inequality (6.20) now follows. ∎

We will use the following notational convention. If $A \subset \mathbb{R}^3$ is a measurable set and $f : A \mapsto \mathbb{R}^3$ is a measurable function, if $\widetilde{f}$ denotes the zero extension of f to $\mathbb{R}^3$, and if $\widetilde{f}$ satisfies the assumptions of Lemma 6.1, we will write $\mathfrak{R}(f)$ instead of $\mathfrak{R}(\widetilde{f})$. A similar convention is to hold with respect to $\mathfrak{S}(g)$ if $g : A \mapsto \mathbb{R}$ is a measurable function such that its zero extension to $\mathbb{R}^3$ verifies the assumptions of Lemma 6.3.

Lemma 6.4 *Let* $R \in (0, \infty)$ *with* $\overline{\mathfrak{D}} \subset B_R$, $f \in L^1(\partial \mathfrak{D}_R)$, $j, k \in \{1, 2, 3\}$, $\alpha \in \mathbb{N}_0^3$ *with* $|\alpha| \leq 1$. *Define*

$$F(y) := \int_{\partial \mathfrak{D}_R} \partial_z^\alpha \mathcal{Z}_{jk}(y, z) f(z) \, do_z, \quad H(y) := \int_{\partial \mathfrak{D}_R} E_{4j}(y - z) f(z) \, do_z$$

for $y \in \mathfrak{D}_R$. *Then* F *and* H *are continuous. Moreover, let* $x \in \mathfrak{D}_R$, *and put* $\delta_x := \mathrm{dist}(\partial \mathfrak{D}_R, x)$. *Then*

$$|F(x)| + |H(x)| \leq \mathfrak{C}(\delta_x, R) \|f\|_1. \tag{6.21}$$

Proof Let $U \subset \mathbb{R}^3$ be open, with $\overline{U} \subset \mathfrak{D}_R$. Then $\delta_U := \mathrm{dist}(\overline{U}, \partial \mathfrak{D}_R) > 0$, so we get by (4.31) that

$$|\partial_z^\alpha \mathcal{Z}_{jk}(y, z) f(z)| \leq \mathfrak{C}(R) \delta_U^{-1-|\alpha|} |f(z)| \quad \text{for } z \in \partial \mathfrak{D}_R.$$

In view of the first statement of Lemma 4.3, we conclude that F is continuous. From (4.31), we get that $|F(x)| \leq \mathfrak{C}(\delta_x, R) \|f\|_1$. Obviously $E_{4j} \in C^\infty(\mathbb{R}^3 \backslash \{0\})$ and $|E_{4j}(x)| \leq |x|^{-2}$ for $x \in \mathbb{R}^3 \backslash \{0\}$, so the function H may be handled in the same way (and even belongs to $C^\infty(\mathfrak{D}_R)$). ∎

Lemma 6.5 *Let* $S \in (0, \infty)$ *with* $\overline{\mathfrak{D}} \subset B_S$. *Let* $f \in L^1(\partial \mathfrak{D})$, $g \in L^1(\mathfrak{D})$, $j, k \in \{1, 2, 3\}$, *and define*

$$F^{(1)}(y) := \int_{\partial \mathfrak{D}} \mathcal{Z}_{jk}(y, z) f(z) \, do_z, \quad F^{(2)}(y) := \int_{\mathfrak{D}} \mathcal{Z}_{jk}(y, z) g(z) \, dz,$$

$$F^{(3)}(y) := \int_{\partial \mathfrak{D}} E_{4j}(y - z) f(z) \, do_z, \quad F^{(4)}(y) := \int_{\mathfrak{D}} \partial_k E_{4j}(y - z) g(z) \, dz$$

for $y \in \overline{\mathfrak{D}}^c$. *Then* $F^{(i)} \in C^1(\overline{\mathfrak{D}}^c)$ *for* $1 \leq i \leq 4$. *Put* $\delta := \mathrm{dist}(\overline{\mathfrak{D}}, \partial B_S)$. *Then*

$$|\partial^\alpha F^{(i)}(y)| \leq \mathfrak{C}(\delta, S) \left(|y| s_\tau(y)\right)^{-1-|\alpha|/2} \|f\|_1, \tag{6.22}$$

$$|\partial^\alpha F^{(j)}(y)| \leq \mathfrak{C}(\delta, S) \left(|y| s_\tau(y)\right)^{-1-|\alpha|/2} \|g\|_1 \tag{6.23}$$

for $y \in B_S^c$, $\alpha \in \mathbb{N}_0^3$ *with* $|\alpha| \leq 1$, $i \in \{1, 3\}$, $j \in \{2, 4\}$.

Proof Let $U \subset \mathbb{R}^3$ be open and bounded with $\overline{U} \subset \overline{\mathfrak{D}}^c$. Let $R \in (0, \infty)$ with $\mathfrak{D} \cup \overline{U} \subset B_R$. Then an argument as in the proof of Lemma 6.4, based on (4.31) and Lemma 4.3, yields that $F^{(1)}|U \in C^1(U)$, and

$$\partial_l F^{(1)}(y) = \int_{\partial \mathfrak{D}} \partial y_l Z_{jk}(y, z) f(z) \, do_z \quad \text{for } y \in U, \ 1 \le l \le 3. \quad (6.24)$$

It follows that $F^{(1)} \in C^1(\overline{\mathfrak{D}}^c)$, and that Eq. (6.24) holds for $y \in \overline{\mathfrak{D}}^c$. Put $S_1 := S - \delta/2$. Then $S_1 \in (0, S)$ and $\overline{\mathfrak{D}} \subset B_{S_1}$, so inequality (4.33), with S, δ replaced by $S_1, S/S_1 - 1$, yields

$$|\partial_y^\alpha Z_{jk}(y, z) f(z)| \le \mathfrak{C}(S, S_1) \big(|y| s_\tau(y)\big)^{-1-|\alpha|/2} |f(z)|$$

for $z \in \partial \mathfrak{D}$, $y \in B_S^c$, $\alpha \in \mathbb{N}_0^3$ with $|\alpha| \le 1$. Now we get by (6.24) that

$$|\partial^\alpha F^{(1)}(y)| \le \mathfrak{C}(\delta, S) \big(|y| s_\tau(y)\big)^{-1-|\alpha|/2} \|f\|_1$$

for y, α as before. The function $F^{(2)}$ may be dealt with in a similar way. As for $F^{(3)}$ and $F^{(4)}$, we note that for $y \in B_S^c$ and $z \in \overline{\mathfrak{D}}$, we have $|y - z| \ge (1 - S_1/S)|y|$. This observation and Lemma 3.8 yield the estimates of $F^{(3)}$ and $F^{(4)}$ stated in (6.22) and (6.23), respectively. ∎

We want to extend Lemmas 6.4, 6.5 to the case when second-order derivatives act on layer and volume potentials.

Lemma 6.6 *Let* $j, k, l \in \{1, 2, 3\}$, $f \in L^1(\partial \mathfrak{D})$, *and put*

$$F(y) := \int_{\partial \mathfrak{D}} \partial z_l Z_{jk}(y, z) \cdot f(z) \, do_z, \quad F^{(1)}(y) := \int_{\partial \mathfrak{D}} Z_{jk}(y, z) \cdot f(z) \, do_z,$$

$$F^{(3)}(y) := \int_{\partial \mathfrak{D}} E_{4j}(y - z) \cdot f(z) \, do_z \quad \text{for } y \in \overline{\mathfrak{D}}^c.$$

Then $F \in C^1(\overline{\mathfrak{D}}^c)$, $F^{(1)}, F^{(3)} \in C^2(\overline{\mathfrak{D}}^c)$, *and*

$$\partial_m F(y) = \int_{\partial \mathfrak{D}} \partial y_m \partial z_l Z_{jk}(y, z) \cdot f(z) \, do_z, \quad (6.25)$$

$$\partial^\alpha F^{(1)}(y) = \int_{\partial \mathfrak{D}} \partial_y^\alpha Z_{jk}(y, z) \cdot f(z) \, do_z, \quad (6.26)$$

$$\partial^\alpha F^{(3)}(y) = \int_{\partial \mathfrak{D}} \partial_y^\alpha E_{4j}(y - z) \cdot f(z) \, do_z \quad (6.27)$$

for $1 \le m \le 3$, $\alpha \in \mathbb{N}_0^3$ *with* $|\alpha| \le 2$, $y \in \overline{\mathfrak{D}}^c$.
   Let $S_1, S \in (0, \infty)$ *with* $\overline{\mathfrak{D}} \subset B_{S_1}$, $S_1 < S$. *Then*

$$|\partial^\beta F(y)| \le \mathfrak{C}(S_1, S) \cdot \|f\|_1 \cdot \left(|y| \cdot s_\tau(y)\right)^{-3/2 - |\beta|/2}, \tag{6.28}$$

$$|\partial^\alpha F^{(1)}(y)| \le \mathfrak{C}(S_1, S) \cdot \|f\|_1 \cdot \left(|y| \cdot s_\tau(y)\right)^{-1 - |\alpha|/2}, \tag{6.29}$$

$$|\partial^\alpha F^{(3)}(y)| \le \mathfrak{C}(S_1, S) \cdot \|f\|_1 \cdot |y|^{-2 - |\alpha|} \tag{6.30}$$

for $y \in B_S^c$, $\alpha, \beta \in \mathbb{N}_0^3$ *with* $|\alpha| \le 2$, $|\beta| \le 1$.

Proof Let $U \subset \mathbb{R}^3$ with $\overline{U} \subset \overline{\mathfrak{D}}^c$. Since $\mathcal{Z}_{jk}$ is a C^2-function on $(\mathbb{R}^3 \times \mathbb{R}^3) \backslash \{(x, x) : x \in \mathbb{R}^3\}$ (Lemma 4.4) and E_{4j} is a C^∞-function on $\mathbb{R}^3 \backslash \{0\}$.

Since $\mathrm{dist}(\partial\mathfrak{D}, \overline{U}) > 0$. it follows from Lebesgue's theorem on dominated convergence that $F_l|U \in C^1(U)$, $F^{(1)}|U$, $F^{(3)}|U \in C^2(U)$, and Eqs. (6.25)–(6.27) hold for $y \in U$. This is true for any $U \subset \mathbb{R}^3$ with $\overline{U} \subset \overline{\mathfrak{D}}^c$, so we have proved that $F \in C^1(\overline{\mathfrak{D}}^c)$, $F^{(1)}, F^{(3)} \in C^2(\overline{\mathfrak{D}}^c)$, and that Eqs. (6.25)–(6.27) hold for $y \in \overline{\mathfrak{D}}^c$.

Inequality (6.28)–(6.30) follow from (6.25)–(6.27), Theorem 4.4 and the relations

$$|\partial^\alpha E_{4j}(x)| \le \mathfrak{C} \cdot |x|^{-2 - |\alpha|} \quad \text{for} \ x \in \mathbb{R}^3 \backslash \{0\}, \ \alpha \in \mathbb{N}_0^3 \text{ with } |\alpha| \le 2,$$

$$|y - z| \ge |y| - |z| = (1 - S_1/S) \cdot |y| + (S_1/S) \cdot |y| - |z| \ge (1 - S_1/S) \cdot |y|. \tag{6.31}$$

∎

The same type of arguments yields

Lemma 6.7 *Let* $j, k \in \{1, 2, 3\}$, $\beta \in \mathbb{N}_0^3$ *with* $|\beta| \le 1$, $R > 0$, $g \in L^1(B_R)$, *and put*

$$F(y) := \int_{B_R} \mathcal{Z}_{jk}(y, z) \cdot g(z) \, dz, \quad G(y) := \int_{B_R} \partial_y^\beta E_{4j}(y - z) \cdot g(z) \, dz \quad \text{for} \ y \in \overline{B_R}^c.$$

Then $F, G \in C^2(\overline{B_R}^c)$ *and*

$$\partial^\alpha F(y) = \int_{B_R} \partial_y^\alpha \mathcal{Z}_{jk}(y, z) \cdot g(z) \, dz, \quad \partial^\alpha G(y) = \int_{B_R} \partial_y^\alpha \partial_y^\beta E_{4j}(y - z) \cdot g(z) \, dz$$

for $\alpha \in \mathbb{N}_0^3$ *with* $|\alpha| \le 2$, $y \in \overline{B_R}^c$. *Let* $T \in (R, \infty)$. *Then*

$$|\partial^\alpha F(y)| \le \mathfrak{C}(R, T) \cdot \|g\|_1 \cdot \left(|y| \cdot s_\tau(y)\right)^{-1 - |\alpha|/2}, \quad |\partial^\alpha G(y)| \le \mathfrak{C}(R, T) \cdot \|g\|_1 \cdot |y|^{-2 - |\alpha| + \beta}$$

for $y \in B_T^c$, $\alpha \in \mathbb{N}_0^3$ *with* $|\alpha| \le 2$.

For convenience of readers we recall representation formula which was derived in the slightly different form in Theorem 5.2

Theorem 6.3 *Let* $R \in (0, \infty)$ *with* $\overline{\mathfrak{D}} \subset \mathbb{R}^3$, *and let* $n^{(R)} : \partial B_R \cup \partial\mathfrak{D} \mapsto \mathbb{R}^3$ *denote the outward unit normal to* $\mathfrak{D}_R$. *Suppose that* $u \in C^2(\overline{\mathfrak{D}_R})^3$, $\pi \in C^1(\overline{\mathfrak{D}_R})$, *and put* $F := \mathcal{L}(u) + \nabla\pi$. *Let* $y \in \mathfrak{D}_R$ *and* $j \in \{1, 2, 3\}$. *Then*

$$u_j(y) = \mathfrak{R}_j(F)(y) + \mathfrak{S}_j(\text{div } u)(y) + \int_{\partial \mathfrak{D}_R} \mathfrak{A}_j^{(R)}(u, \pi)(y, z) \, do_z, \quad (6.32)$$

where

$$\mathfrak{A}_j^{(R)}(u, \pi)(y, z) \quad\quad\quad\quad\quad\quad\quad\quad\quad\quad\quad\quad\quad\quad\quad (6.33)$$

$$:= \sum_{k=1}^{3} \Big[\sum_{l=1}^{3} \Big(\mathcal{Z}_{jk}(y, z) \big(\partial_l u_k(z) - \delta_{kl}\pi(z) + u_k(z)(-\tau e_1 + \omega \times z)_l \big)$$

$$- \partial_{z_l} \mathcal{Z}_{jk}(y, z) u_k(z) \Big) n_l^{(R)}(z) - E_{4j}(y - z) u_k(z) n_k^{(R)}(z) \Big]$$

for $y \in \mathfrak{D}_R$, $z \in \partial \mathfrak{D}_R$.

Outline of proof: Let $\epsilon \in (0, \infty)$ with $\overline{B_\epsilon(y)} \subset \mathfrak{D}_R$, and consider the integral

$$A_{j,\epsilon} := \int_{\mathfrak{D}_R \backslash B_\epsilon(y)} \sum_{k=1}^{3} \mathcal{Z}_{jk}(y, z) \big(\mathcal{L}(u) + \nabla\pi \big)_k(z) \, dz.$$

By performing some integrations by parts, using (3.24), integrating with respect to t, and then exploiting (3.26), we obtain

$$A_{j,\epsilon} = \int_{\mathfrak{D}_R \backslash B_\epsilon(y)} -E_{4j}(y - z) \text{div } u(z) \, dz - S_{j,\epsilon}(y),$$

where $S_{j,\epsilon}(y)$ denotes a surface integral defined in the same way as the surface integral on the right-hand side of (6.32), but with $\partial B_R \cup \partial \mathfrak{D} \cup \partial B_\epsilon(y)$ as domain of integration instead of $\partial B_R \cup \partial \mathfrak{D}$, and with $n^{(R)}$ replaced by the outward unit normal to $\mathfrak{D}_R \backslash B_\epsilon(y)$. Equation (6.32) then follows by a passage to the limit $\epsilon \downarrow 0$, with the calculation of $\lim_{\epsilon \downarrow 0} S_{j,\epsilon}(y)$ based on Theorem 5.1. This reasoning requires some applications of Fubini's and Lebesgue's theorem, all of which made possible by Lemma 4.2. ∎

Our next aim consists in extending Eq. (6.32) to functions u and π which are less regular than C^2 and C^1, respectively. We begin by specifying the type of functions we will consider. From now we need that $\partial \mathfrak{D}$ is of class C^2. (Theorem 6.3 also holds if $\mathfrak{D}$ is only Lipschitz bounded.)

Definition 6.1 Let $p \in (1, \infty)$. Define $\mathfrak{M}_p$ as the space of all pairs of functions such that $u \in W_{\text{loc}}^{2,p}(\overline{\mathfrak{D}}^c)^3$, $\pi \in W_{\text{loc}}^{1,p}(\overline{\mathfrak{D}}^c)$,

$$u|\mathfrak{D}_T \in W^{1,p}(\mathfrak{D}_T)^3, \quad \pi|\mathfrak{D}_T \in L^p(\mathfrak{D}_T), \quad u|\partial \mathfrak{D} \in W^{2-1/p, p}(\partial \mathfrak{D})^3, \quad (6.34)$$

$$\text{div } u|\mathfrak{D}_T \in W^{1,p}(\mathfrak{D}_T), \quad L(u) + \nabla\pi|\mathfrak{D}_T \in L^p(\mathfrak{D}_T)^3$$

for some $T \in (0, \infty)$ with $\overline{\mathfrak{D}} \subset B_T$.

Theorem 6.4 *Let $p \in (1, \infty)$, $(u, \pi) \in \mathfrak{M}_p$.*
Then $u|\mathfrak{D}_T \in W^{2,p}(\mathfrak{D}_T)^3$, $\pi|\mathfrak{D}_T \in W^{1,p}(\mathfrak{D}_T)$ for any $T \in (0, \infty)$ with $\overline{\mathfrak{D}} \subset B_T$.

Proof The theorem follows from the regularity theory for the Stokes system. To be more specific, we first note that our assumptions imply that the relations in (6.34) hold for all $T \in (0, \infty)$ with $\overline{\mathfrak{D}} \subset B_T$. Take such a number T. Let $S \in (T, \infty)$, and choose $\zeta \in C_0^\infty(\mathbb{R}^3)$ with $\zeta|B_T = 1$, $\zeta|B_S^c = 0$. Then

$$\zeta u \,|\, \mathfrak{D}_S \in W_{\text{loc}}^{2,p}(\mathfrak{D}_S)^3 \cap W^{1,p}(\mathfrak{D}_S)^3, \quad \zeta \pi \,|\, \mathfrak{D}_S \in W_{\text{loc}}^{1,p}(\mathfrak{D}_S) \cap L^p(\mathfrak{D}_S), \qquad (6.35)$$

$$\text{div } (\zeta u) \,|\, \mathfrak{D}_S \in W^{1,p}(\mathfrak{D}_S),$$

$$\zeta u \,|\, \partial \mathfrak{D} = u|\partial \mathfrak{D} \in W^{2-1/p,p}(\partial \mathfrak{D})^3, \quad \text{hence} \quad \zeta u \,|\, \partial \mathfrak{D}_S \in W^{2-1/p,p}(\partial \mathfrak{D}_S)^3.$$

Moreover, since $u|\mathfrak{D}_S \in W^{1,p}(\mathfrak{D}_S)^3$, $L(u) + \nabla \pi \,|\, \mathfrak{D}_S \in L^p(\mathfrak{D}_S)^3$, we have $-\Delta u + \nabla \pi \,|\, \mathfrak{D}_S \in L^p(\mathfrak{D}_S)^3$. Once more observing that $u|\mathfrak{D}_S \in W^{1,p}(\mathfrak{D}_S)^3$, $\pi|\mathfrak{D}_S \in L^p(\mathfrak{D}_S)$, we may conclude that

$$-\Delta(\zeta u) + \nabla(\zeta \pi) \,|\, \mathfrak{D}_S \in L^p(\mathfrak{D}_S)^3. \qquad (6.36)$$

Obviously the function ζu is a weak solution of the Stokes system in $\mathfrak{D}_S$ with right-hand side $-\Delta(\zeta u) + \nabla(\zeta \pi) \,|\, \mathfrak{D}_S$, where "weak solution" is meant in the sense of [27, (IV.1.3)]. In view of (6.35) and (6.36), it follows from [27, Lemma IV.6.1, Exercise IV.6.2] that $\zeta u \,|\, \mathfrak{D}_S \in W^{2,p}(\mathfrak{D}_S)^3$, $\zeta \pi \,|\, \mathfrak{D}_S \in W^{1,p}(\mathfrak{D}_S)$. This implies that $u|\mathfrak{D}_T \in W^{2,p}(\mathfrak{D}_T)^3$ and $\pi|\mathfrak{D}_T \in W^{1,p}(\mathfrak{D}_T)$. ∎

Now we are in a position to generalize Theorem 6.3 to pairs of functions $(u, \pi) \in \mathfrak{M}_p$.

Theorem 6.5 *Let $p \in (1, \infty)$, $(u, \pi) \in \mathfrak{M}_p$, $j \in \{1, 2, 3\}$. Put $F := \mathcal{L}(u) + \nabla \pi$.*
Take R and $n^{(R)}$ as in Theorem 6.3. Then, for a.e. $y \in \mathfrak{D}_R$,

$$u_j(y) = \mathfrak{R}_j(F|\mathfrak{D}_R)(y) + \mathfrak{S}_j(\text{div } u|\mathfrak{D}_R)(y) + \int_{\partial \mathfrak{D}_R} \mathfrak{A}_j^{(R)}(u, \pi)(y, z) \, do_z, \qquad (6.37)$$

with $\mathfrak{A}_j^{(R)}(u, \pi)(y, z)$ defined as in (6.33).
If $p > 3/2$, Eq. (6.37) holds for any $y \in \mathfrak{D}_R$ (without the restriction "a.e.").

Proof By Theorem 6.4, we have $u|\mathfrak{D}_R \in W^{2,p}(\mathfrak{D}_R)^3$ and $\pi|\mathfrak{D}_R \in W^{1,p}(\mathfrak{D}_R)$. Therefore (see [1, (3.18)]) there are sequences (u_n) in $C^\infty(\mathbb{R}^3)^3$ and (π_n) in $C^\infty(\mathbb{R}^3)$ with

$$\|(u - u_n)|\mathfrak{D}_R\|_{2,p} + \|(\pi - \pi_n)|\mathfrak{D}_R\|_{1,p} \to 0. \qquad (6.38)$$

By a standard trace theorem, it follows that $u_k|\partial \mathfrak{D}_R$, $\partial_l u_k|\partial \mathfrak{D}_R$ and $\pi|\partial \mathfrak{D}_R$ belong to $L^1(\partial \mathfrak{D}_R)$, and

$$\|(u - u_n)|\partial \mathfrak{D}_R\|_1 + \|(\partial_l u - \partial_l u_n)|\partial \mathfrak{D}_R\|_1 + \|(\pi - \pi_n)|\partial \mathfrak{D}_R\|_1 \to 0 \qquad (6.39)$$

for $n \to \infty$ $(1 \le k, l \le 3)$. Let $y \in \mathfrak{D}_R$. We may conclude from (6.21) and (6.39) that

$$\int_{\partial \mathfrak{D}_R} \mathfrak{A}_j^{(R)}(u_n, \pi_n)(y, z) \, do_z \to \int_{\partial \mathfrak{D}_R} \mathfrak{A}_j^{(R)}(u, \pi)(y, z) \, do_z \quad (n \to \infty), \quad (6.40)$$

where the definition of $\mathfrak{A}_j^{(R)}(u_n, \pi_n)(y, z)$ should be obvious by (6.33). For $n \in \mathbb{N}$, we set $F_n := L(u_n) + \nabla \pi_n$. By (6.38), we have

$$\|(F_n - F)|\mathfrak{D}_R\|_p \to 0, \quad \|\operatorname{div}(u - u_n)|\mathfrak{D}_R\|_p \to 0 \quad (n \to \infty).$$

These relations combined with (6.3) and (6.18) imply

$$\|\mathfrak{R}_j\big((F_n - F)|\mathfrak{D}_R\big) | \mathfrak{D}_R\|_p + \|\mathfrak{S}_j\big(\operatorname{div}(u_n - u)|\mathfrak{D}_R\big) | \mathfrak{D}_R\|_p \to 0 \quad (n \to \infty).$$

Passing from L^p-convergence to pointwise convergence of subsequences and recalling (6.38), we see that there is a strictly increasing function $\sigma : \mathbb{N} \mapsto \mathbb{N}$ such that

$$\mathfrak{R}_j \ (F_{\sigma(n)}|\mathfrak{D}_R)(y) \to \mathfrak{R}_j(F|\mathfrak{D}_R)(y), \quad (6.41)$$
$$\mathfrak{S}_j(\operatorname{div} u_{\sigma(n)}|\mathfrak{D}_R)(y) \to \mathfrak{S}_j(\operatorname{div} u|\mathfrak{D}_R)(y), \quad u_{\sigma(n)}(y) \to u(y) \quad (n \to \infty)$$

for a. e. $y \in \mathfrak{D}_R$. On the other hand, by Theorem 6.3, Eq. (6.37) holds with u, π replaced by u_n, π_n, respectively, for $n \in \mathbb{N}$. Therefore we may conclude from (6.40) and (6.41) that Eq. (6.37) holds for a. e. $y \in \mathfrak{D}_R$.

Now suppose that $p > 3/2$. Since $(u, \pi) \in \mathfrak{M}_p$ and because of the Sobolev inequality (in the case $p \le 3$), we may conclude that $\operatorname{div} u|\mathfrak{D}_R \in L^q(\mathfrak{D}_R)$ for some $q \in (3, \infty)$. Recalling the relation $F|\mathfrak{D}_R \in L^p(\mathfrak{D}_R)^3$, we thus see by Lemmas 6.2, 6.3 with $S = R$ that $\mathfrak{R}(F|\mathfrak{D}_R)$ and $\mathfrak{S}(\operatorname{div} u|\mathfrak{D}_R)$ are continuous. Moreover, since $p > 3/2$ and $u|\mathfrak{D}_R \in W^{2,p}(\mathfrak{D}_R)^3$, the Sobolev lemma implies that u may be considered as a continuous function on $\overline{\mathfrak{D}}_R$. According to Lemma 6.4, the function associating the integral $\int_{\partial \mathfrak{D}_R} \mathfrak{A}_j^{(R)}(y, z) \, do_z$ with each $y \in \mathfrak{D}_R$ is also continuous. Thus we may conclude that Eq. (6.37) is valid for any $y \in \mathfrak{D}_R$, without the restriction "a.e.". $\blacksquare$

Next we perform the transition from a representation formula on $\mathfrak{D}_R$ to one on $\overline{\mathfrak{D}}^c$. For this step, we only need the decay properties given implicitly by the relations in (6.42).

Theorem 6.6 *Let* $p \in (1, \infty)$, $(u, \pi) \in \mathfrak{M}_p$. *Put* $F := \mathcal{L}(u) + \nabla \pi$, *and suppose there are numbers* $p_1, p_2 \in (1, 2)$, $S \in (0, \infty)$ *such that* $\overline{\mathfrak{D}} \subset B_S$,

$$u|B_S^c \in L^6(B_S^c)^3, \quad \nabla u|B_S^c \in L^2(B_S^c)^9 \ , \quad \pi|B_S^c \in L^2(B_S^c), \quad (6.42)$$
$$F|B_S^c \in L^{p_1}(B_S^c)^3 + L^{p_2}(B_S^c)^3.$$

Let $j \in \{1, 2, 3\}$, and put

$$\mathfrak{B}_j(y) := \mathfrak{B}_j(u, \pi)(y) \tag{6.43}$$

$$:= \int_{\partial\mathfrak{D}} \sum_{k=1}^{3} \Big[\sum_{l=1}^{3} \Big(\mathcal{Z}_{jk}(y, z) \big(-\partial_l u_k(z) + \delta_{kl}\,\pi(z) + u_k(z)(\tau e_1 - \omega \times z)_l \big)$$

$$+ \partial_{z_l} \mathcal{Z}_{jk}(y, z) u_k(z) \Big) n_l^{(\mathfrak{D})}(z) \; + \; E_{4j}(y - z) u_k(z) n_k^{(\mathfrak{D})}(z) \Big]\, do_z$$

for $y \in \overline{\mathfrak{D}}^c$. Then

$$u_j(y) = \mathfrak{R}_j(F)(y) + \mathfrak{S}_j(div\,u)(y) + \mathfrak{B}_j(y) \tag{6.44}$$

for a.e. $y \in \overline{\mathfrak{D}}^c$. If $p > 3/2$, Eq. (6.44) holds for any $y \in \overline{\mathfrak{D}}^c$, without the restriction "a.e.".

Proof The assumptions on u and π yield that

$$\int_{S}^{\infty} \int_{\partial B_r} \big(|u(z)|^6 + |\nabla u(z)|^2 + |\pi(z)|^2 \big)\, do_z\, dr < \infty. \tag{6.45}$$

Therefore there is an increasing sequence (R_n) in (S, ∞) with $R_n \to \infty$ and

$$\int_{\partial B_{R_n}} \big(|u(z)|^6 + |\nabla u(z)|^2 + |\pi(z)|^2 \big)\, do_z \le R_n^{-1} \quad \text{for } n \in \mathbb{N}. \tag{6.46}$$

Otherwise there would be a constant $C \in [S, \infty)$ such that

$$\int_{\partial B_r} \big(|u(z)|^6 + |\nabla u(z)|^2 + |\pi(z)|^2 \big)\, do_z \ge r^{-1} \quad \text{for } r \in [C, \infty),$$

in contradiction to (6.45). (Here we have used a standard convention from the theory of Lebesgue integration, which states that the integral of every measurable nonnegative function is defined, but may take the value ∞.) By our assumptions on F, there are functions $G^{(i)} \in L^{p_i}(B_S^c)^3$ for $i \in \{1, 2\}$ such that $F|B_S^c = G^{(1)} + G^{(2)}$. Thus, by Lemma 6.1,

$$\int_{\mathbb{R}^3} \sum_{k=1}^{3} |\mathcal{Z}_{jk}(y, z)| \Big(\chi_{(0,S]}(|z|)\, |F_k(z)| \tag{6.47}$$

$$+ \chi_{(S,\infty)}(|z|) \big(|G_k^{(1)}(z)| + |G_k^{(2)}(z)| \big) \Big)\, dz < \infty$$

for a. e. $y \in \overline{\mathfrak{D}}^c$. Moreover, by Lemma 6.3 with $q = 2$,

$$\int_{\mathbb{R}^3} |E_{4j}(y - z)| \, |\operatorname{div} u(z)| \, dz < \infty \qquad (6.48)$$

for a. e. $y \in \overline{\mathfrak{D}}^c$. Due to these observations and Theorem 6.5, we see there is a subset N of $\overline{\mathfrak{D}}^c$ with measure zero such that the relations in (6.47) and (6.48) hold for $y \in \overline{\mathfrak{D}}^c \backslash N$, and such that Eq. (6.37) with R replaced by R_n holds for $n \in \mathbb{N}$ and $y \in \mathfrak{D}_R \backslash N$. In the case $p > 3/2$, Lemma 6.2 yields that (6.47) is valid for any $y \in \overline{\mathfrak{D}}^c$, and Theorem 6.5 implies that Eq. (6.37) with R replaced by R_n is true for $n \in \mathbb{N}$ and any $y \in \overline{\mathfrak{D}}^c$. Moreover, if $p > 3/2$, the assumption $(u, \pi) \in \mathfrak{M}_p$, Lemma 6.3 and the Sobolev inequality (in the case $p \leq 3$) allow to drop the restriction "a. e." in (6.48).

Take $y \in \overline{\mathfrak{D}}^c$ in the case $p > 3/2$ and $y \in \overline{\mathfrak{D}}^c \backslash N$ otherwise. Let $n \in \mathbb{N}$ with $R_n > |y|$ (hence $y \in \mathfrak{D}_{R_n}$). Then, by Eq. (6.37) with R replaced by R_n, we get

$$u_j(y) = \mathfrak{R}_j(F|\mathfrak{D}_{R_n})(y) + \mathfrak{S}_j(\operatorname{div} u|\mathfrak{D}_{R_n})(y) + \mathfrak{A}_{j,n}(y) + \mathfrak{B}_j(y), \quad (6.49)$$

with

$$\mathfrak{A}_{j,n}(y) := \int_{\partial B_{R_n}} \sum_{k=1}^{3} \Big[\sum_{l=1}^{3} \Big(\mathcal{Z}_{jk}(y, z) \big(\partial_l u_k(z) - \delta_{kl}\pi(z) - \tau \delta_{1l} u_k(z) \big)$$
$$- \partial_{z_l} \mathcal{Z}_{jk}(y, z) u_k(z) \Big) z_l / R_n \; - \; E_{4j}(y - z) u_k(z) z_k / R_n \Big] \, do_z.$$

Note that in (6.49), we used the relation $\sum_{l=1}^{3}(\omega \times z)_l z_l / R_n = 0$ for $z \in \partial B_R$. The term $\mathfrak{B}_j(y)$ was defined in (6.43). Let $n \in \mathbb{N}$ with $R_n/4 \geq |y|$. Observe that

$$|\mathfrak{A}_{j,n}(y)| \leq \mathfrak{C} \sum_{\nu=1}^{4} \sum_{k=1}^{3} \mathfrak{V}_{\nu,k}(y), \qquad (6.50)$$

with

$$\mathfrak{V}_{1,k}(y) := \Big(\int_{\partial B_{R_n}} |\mathcal{Z}_{jk}(y, z)|^{6/5} \, do_z \Big)^{5/6} \|u|\partial B_{R_n}\|_6,$$

$$\mathfrak{V}_{2,k}(y) := \Big(\int_{\partial B_{R_n}} |\mathcal{Z}_{jk}(y, z)|^2 \, do_z \Big)^{1/2} (\|\nabla u|\partial B_{R_n}\|_2 + \|\pi|\partial B_{R_n}\|_2),$$

$$\mathfrak{V}_{3,k}(y) := \sum_{l=1}^{3} \Big(\int_{\partial B_{R_n}} |\partial_{z_l} \mathcal{Z}_{jk}(y, z)|^{6/5} \, do_z \Big)^{5/6} \|u|\partial B_{R_n}\|_6,$$

$$\mathfrak{V}_{4,k}(y) := \Big(\int_{\partial B_{R_n}} |y - z|^{-12/5} \, do_z \Big)^{5/6} \|u|\partial B_{R_n}\|_6$$

for $k \in \{1, 2, 3\}$. Since $|y| \leq R_n/4$, we may use inequality (4.34) with $S = 2|y|$ in order to estimate $|\partial_z^\alpha \mathcal{Z}_{jk}(y, z)|$ for $z \in \partial B_{R_n}$, $\alpha \in \mathbb{N}_0^3$ with $|\alpha| \leq 1$. We get by (6.46) and (4.34) that

$$\mathfrak{V}_{1,k}(y) \leq \mathfrak{C}(|y|) \left(\int_{\partial B_{R_n}} \left(|z| s_\tau(z) \right)^{-6/5} do_z \right)^{5/6} R_n^{-1/6} \tag{6.51}$$

$$\leq \mathfrak{C}(|y|) \left(\int_{\partial B_{R_n}} s_\tau(z)^{-6/5} do_z \right)^{5/6} R_n^{-7/6} \leq \mathfrak{C}(|y|) R_n^{-1/3},$$

where the last inequality follows from Lemma 3.5. The same references yield

$$|\mathfrak{V}_{2,k}(y)| \leq \mathfrak{C}(|y|) R_n^{-1}, \quad |\mathfrak{V}_{3,k}(y)| \leq \mathfrak{C}(|y|) R_n^{-5/6} \quad (1 \leq k \leq 3). \tag{6.52}$$

Moreover, since $|y - z| \geq |z|/2$ for ∂B_{R_n}, we find with (6.46) that $|\mathfrak{V}_{4,k}(y)| \leq \mathfrak{C}(|y|) R_n^{-1/2}$. From (6.50)–(6.52) and the preceding inequality we may conclude that $\mathfrak{A}_{n,j}(y) \to 0$ for $n \to \infty$. Turning to $\mathfrak{R}_j(F|\mathfrak{D}_{R_n})(y)$, we observe that by (6.47), our choice of y and Lebesgue's theorem on dominated convergence, we have $\mathfrak{R}_j(F|\mathfrak{D}_{R_n})(y) \to \mathfrak{R}_j(F)(y)$ for $n \to \infty$. Moreover, by (6.48) and again by the choice of y and Lebesgue's theorem, $\mathfrak{S}_j(\operatorname{div} u|\mathfrak{D}_{R_n})(y) \to \mathfrak{S}_j(\operatorname{div} u)(y)$ for $n \to \infty$. Recalling (6.49), we thus have proved (6.44). ∎

In our first application of our representation formula (6.44), we state conditions on $\mathcal{L}(u) + \nabla \pi$ and $\operatorname{div} u$ such that u decays as described

$$|u(x)| = O\left[\left(|x| \left(1 + \tau(|x| - x_1) \right) \right)^{-1} \right], \tag{6.53}$$

$$|\nabla u(x)| = O\left[\left(|x| \left(1 + \tau(|x| - x_1) \right) \right)^{-3/2} \right] \quad \text{for } |x| \to \infty.$$

Since in the proof of this result, we want to avoid estimates of the second derivatives of $\mathcal{Z}_{jk}$, we have to transform the integral $\int_{\partial \mathfrak{D}} \partial_{z_l} \mathcal{Z}_{jk}(y, z) u_k(z) n_l^{(\mathfrak{D})}(z) \, do_z$ appearing in the definition of $\mathfrak{B}_j(y)$ (see (6.43)), into a term where no differential operator acts on $\mathcal{Z}_{jk}$. This is done in

Lemma 6.8 *Let $p \in (1, \infty)$, $(u, \pi) \in \mathfrak{M}_p$, $j \in \{1, 2, 3\}$. Define*

$$\mathfrak{U}_j(y) := \mathfrak{U}_j(u)(y) := \int_{\partial \mathfrak{D}} \sum_{k,l=1}^3 \partial_{z_l} \mathcal{Z}_{jk}(y, z) u_k(z) n_l^{(\mathfrak{D})}(z) \, do_z \tag{6.54}$$

for $y \in \overline{\mathfrak{D}}^c$. Let $\mathcal{E}_p : W^{2-1/p,p}(\partial \mathfrak{D}) \mapsto W^{2,p}(\mathfrak{D})$ denote a continuous extension operator [51]. Then, for $y \in \overline{\mathfrak{D}}^c$,

$$\mathfrak{U}_j(y) = \int_{\mathfrak{D}} \sum_{k=1}^{3} \left[\partial_k E_{4j}(y-z)\, \mathcal{E}_p(u_k)(z) \; + \; \mathcal{Z}_{jk}(y,z) \right. \tag{6.55}$$

$$\left((\tau e_1 - \omega \times z) \cdot \nabla \mathcal{E}_p(u_k)(z) + \left[\omega \times (\mathcal{E}_p(u_s)(z))_{1\le s \le 3} \right]_k - \Delta \mathcal{E}_p(u_k)(z) \right) \Big] dz$$

$$+ \int_{\partial \mathfrak{D}} \sum_{k,l=1}^{3} \mathcal{Z}_{jk}(y,z) \left((-\tau e_1 + \omega \times z)_l\, u_k(z) + \partial_l \mathcal{E}_p(u_k)(z) \right) n_l^{(\mathfrak{D})}(z)\, do_z.$$

Proof Let $y \in \overline{\mathfrak{D}}^c$. Starting with (4.30), we may refer to Lemma 4.2 in order to apply Fubini's theorem and Lebesgue's theorem on dominated convergence, to obtain

$$\mathfrak{U}_j(y) = \lim_{\delta \downarrow 0,\, T \to \infty} \int_{\delta}^{T} \int_{\partial \mathfrak{D}} \sum_{k,l=1}^{3} \partial_{z_l} \Gamma_{jk}(y,z,t)\, u_k(z)\, n_l^{(\mathfrak{D})}(z)\, do_z\, dt.$$

Next we apply the Divergence theorem and then use (3.24). It follows

$$\mathfrak{U}_j(y) = \lim_{\delta \downarrow 0,\, T \to \infty} \int_{\delta}^{T} \int_{\mathfrak{D}} \sum_{k=1}^{3} \Big(\Delta_z \Gamma_{jk}(y,z,t)\, \mathcal{E}_p(u_k)(z) \tag{6.56}$$

$$+ \nabla_z \Gamma(y,z,t) \cdot \nabla \mathcal{E}_p(u_k)(z) \Big)\, dz\, dt$$

$$= \lim_{\delta \downarrow 0,\, T \to \infty} \left[\int_{\delta}^{T} \int_{\mathfrak{D}} \sum_{k=1}^{3} \Big(\big(\partial_t \Gamma_{jk}(y,z,t) + (-\tau e_1 + \omega \times z) \cdot \nabla_z \Gamma_{jk}(y,z,t) \right.$$

$$- \left[\omega \times (\Gamma_{js}(y,z,t))_{1 \le s \le 3} \right]_k \Big) \mathcal{E}_p(u_k)(z) \; - \; \Gamma_{jk}(y,z,t)\, \Delta \mathcal{E}_p(u_k)(z) \Big)\, dz\, dt$$

$$+ \int_{\delta}^{T} \int_{\partial \mathfrak{D}} \sum_{k,l=1}^{3} \Gamma_{jk}(y,z,t)\, \partial_l \mathcal{E}_p(u_k)(z)\, n_l^{(\mathfrak{D})}(z)\, do_z\, dt \bigg].$$

As explained in the proof of Theorem 5.2, the relation in (3.26) and Lemma 4.2 yield

$$\lim_{\delta \downarrow 0,\, T \to \infty} \int_{\delta}^{T} \int_{\mathfrak{D}} \sum_{k=1}^{3} \partial_t \Gamma_{jk}(y,z,t)\, \mathcal{E}_p(u_k)(z)\, dz\, dt \tag{6.57}$$

$$= \int_{\mathfrak{D}} \sum_{k=1}^{3} \partial_k E_{4j}(y-z)\, \mathcal{E}_p(u_k)(z)\, dz.$$

For the other terms on the right-hand side of (6.56), the passage to the limit $\delta \downarrow 0$ and $T \to \infty$ presents no difficulty because due to Lemma 4.2, we may directly apply Fubini's and Lebesgue's theorem. We further use the formula $(a \times b) \cdot c = -(a \times c) \cdot b$ for vectors a, b, c in $\mathbb{R}^3$. In this way, letting δ tend to zero and T to infinity, and taking account of (6.57), we may deduce (6.55) from (6.56). ∎

Now we may prove a decay estimate for $\mathfrak{B}_j(u,\pi)$.

Lemma 6.9 *Let* $p \in (1, \infty)$, $(u, \pi) \in \mathfrak{M}_p$, $j \in \{1, 2, 3\}$. *Define* $\mathfrak{B}_j = \mathfrak{B}_j(u, \pi)$ *as in (6.43). Then* $\mathfrak{B}_j \in C^1(\overline{\mathfrak{D}}^c)$.

Let $S \in (0, \infty)$ *with* $\overline{\mathfrak{D}} \subset B_S$. *Put* $\delta := \mathrm{dist}(\overline{\mathfrak{D}}, \partial B_S)$. *Let* $\alpha \in \mathbb{N}_0^3$ *with* $|\alpha| \leq 1$, $y \in B_S^c$. *Then*

$$|\partial^\alpha \mathfrak{B}_j(y)| \tag{6.58}$$
$$\leq \mathfrak{C}(S, \delta) \big(\|\nabla u \,|\, \partial\mathfrak{D}\|_1 + \|\pi|\partial\mathfrak{D}\|_1 + \widetilde{C}(\mathfrak{D}, p) \, \|u|\partial\mathfrak{D}\|_{2-1/p, p} \big)$$
$$\big(|y| s_\tau(y) \big)^{-1-|\alpha|/2},$$

where $\widetilde{C}(\mathfrak{D}, p)$ *is a constant depending only on* $\mathfrak{D}$ *and* p.

Proof We use the decomposition $\mathfrak{B}_j(y) = \big(\mathfrak{B}_j(y) - \mathfrak{U}_j(y) \big) + \mathfrak{U}_j(y)$, with $\mathfrak{U}_j = \mathfrak{U}_j(u, \pi)$ defined in (6.54). Equation (6.55) and Lemma 6.5 yield that $\mathfrak{B}_j - \mathfrak{U}_j$ and $\mathfrak{U}_j$ belong to $C^1(\overline{\mathfrak{D}}^c)$. Therefore we have $\mathfrak{B}_j \in C^1(\overline{\mathfrak{D}}^c)$. Moreover, by (6.22), (6.23), (6.43) and (6.55),

$$|\partial^\alpha(\mathfrak{B}_j - \mathfrak{U}_j)(y)| + |\partial^\alpha \mathfrak{U}_j(y)| \tag{6.59}$$
$$\leq \mathfrak{C}(S, \delta) \big(|y| s_\tau(y) \big)^{-1-|\alpha|/2} \Big(\|\nabla u \,|\, \partial\mathfrak{D}\|_1 + \|\pi|\partial\mathfrak{D}\|_1 + \|u|\partial\mathfrak{D}\|_1$$
$$+ \sum_{k=1}^{3} \big(\|\mathcal{E}_p(u_k)\|_{2,1} + \|\nabla\mathcal{E}_p(u_k) \,|\, \partial\mathfrak{D}\|_1 \big) \Big),$$

where the extension operator $\mathcal{E}_p$ was introduced in Lemma 6.8. On the other hand, by a standard trace theorem and by the choice of $\mathcal{E}_p$,

$$\|\nabla\mathcal{E}_p(u_k) \,|\, \partial\mathfrak{D}\|_1 \leq \mathfrak{C} \|\nabla\mathcal{E}_p(u_k) \,|\, \partial\mathfrak{D}\|_p \leq \mathfrak{C}(p) \|\mathcal{E}_p(u_k)\|_{2,p} \tag{6.60}$$
$$\leq \mathfrak{C}(p) \|u|\partial\mathfrak{D}\|_{2-1/p, p},$$
$$\|\mathcal{E}_p(u_k)\|_{2,1} \leq \mathfrak{C} \|\mathcal{E}_p(u_k)\|_{2,p} \leq \mathfrak{C}(p) \|u|\partial\mathfrak{D}\|_{2-1/p, p} \tag{6.61}$$

for $k \in \{1, 2, 3\}$. Inequality (6.58) is a consequence of (6.59)–(6.61). ∎

At this point, we are in a position to derive the decay relations (6.53) for u if $\mathcal{L}(u) + \nabla\pi$ and $\mathrm{div}\, u$ decay sufficiently fast.

Theorem 6.7 *Let* $p \in (1, \infty)$, $(u, \pi) \in \mathfrak{M}_p$. *Put* $F := L(u) + \nabla\pi$. *Suppose there are numbers* $S_1, S, \gamma \in (0, \infty)$, $A \in [2, \infty)$, $B \in \mathbb{R}$ *such that* $S_1 < S$, $\overline{\mathfrak{D}} \subset B_{S_1}$,

$$u|B_S^c \in L^6(B_S^c)^3, \quad \nabla u|B_S^c \in L^2(B_S^c)^9, \quad \pi|B_S^c \in L^2(B_S^c), \quad \mathrm{supp}(div\, u) \subset B_{S_1},$$
$$A + \min\{1, B\} \geq 3, \quad |F(z)| \leq \gamma |z|^{-A} s_\tau(z)^{-B} \text{ for } z \in B_{S_1}^c.$$

Put $\delta := \mathrm{dist}(\overline{\mathfrak{D}}, \partial B_S)$. *Let* $i, j \in \{1, 2, 3\}$, $y \in B_S^c$. *Then*

$$|u_j(y)| \le \mathfrak{C}(S, S_1, A, B, \delta)\left(\gamma + \|F|B_{S_1}\|_1 + \|div\, u\|_1 + \|\nabla u\,|\,\partial\mathfrak{D}\|_1 \right.\quad (6.62)$$

$$\left. + \|\pi|\partial\mathfrak{D}\|_1 + \tilde{C}(\mathfrak{D}, p)\|u|\partial\mathfrak{D}\|_{2-1/p,p}\right)\left(|y|s_\tau(y)\right)^{-1} l_{A,B}(y), (6.63)$$

$$|\partial_i u_j(y)|$$

$$\le \mathfrak{C}(S, S_1, A, B, \delta)\left(\gamma + \|F|B_{S_1}\|_1 + \|div\, u\|_1 + \|\nabla u\,|\,\partial\mathfrak{D}\|_1 + \|\pi|\partial\mathfrak{D}\|_1\right.$$

$$\left. + \tilde{C}(\mathfrak{D}, p)\|u|\partial\mathfrak{D}\|_{2-1/p,p}\right)\left(|y|s_\tau(y)\right)^{-3/2} s_\tau(y)^{\max(0,\,7/2-A-B)} l_{A,B}(y),$$

where $\tilde{C}(\mathfrak{D}, p)$ was introduced in Lemma 6.9 and function $l_{A,B}(y)$ in Theorem 6.1. If the assumption $\mathrm{supp}(div\, u) \subset B_{S_1}$ is replaced by the condition

$$|div\, u(z)| \le \tilde{\gamma}|z|^{-C} s_\tau(z)^{-D} \quad for\ \ z \in B_{S_1}^c,$$

for some $\tilde{\gamma} \in (0, \infty)$, $C \in (5/2, \infty)$, $D \in \mathbb{R}$ with $C + \min\{1, D\} > 3$, then inequality (6.62) remains valid if the term $\|div\, u\|_1$ on the right-hand side of (6.62) is replaced by $\tilde{\gamma} + \|div\, u|B_{S_1}\|_1$. Of course, in that case the constant in (6.62) additionally depends on C and D.

Note that if $A + \min\{1, B\} > 3$, $A + B \ge 7/2$ in Theorem 6.7, then $l_{A,B}(y) = 1$ in (6.62) and $s_\tau(y)^{\max(0,\,7/2-A-B)} l_{A,B}(y) = 1$ in (6.64). The preceding conditions on A and B are verified if for example $A = 5/2$, $B = 1$, or $B = 3/2$ and $A = 2 + \epsilon$ for some $\epsilon \in (0, 1/2)$.

Proof of Theorem 6.7: By Lemma 3.5, we see that $\int_{B_{S_1}^c} |F(z)|^r\, dz < \infty$ for any $r \in (1, \infty)$. Thus Theorem 6.6 yields that the representation formula (6.44) holds for a. e. $y \in \overline{\mathfrak{D}}^c$. Therefore Theorem 6.7 follows from Theorems 6.1, 6.2 and Lemma 6.9. ∎

In the next theorem, we present an asymptotic profile of u for the case that $\mathcal{L}(u) + \nabla\pi$ and $div\, u$ have compact support.

Theorem 6.8 *Let $p \in (1, \infty)$, $(u, \pi) \in \mathfrak{M}_p$, $S, S_1 \in (0, \infty)$ with $S_1 < S$, and put $F := \mathcal{L}(u) + \nabla\pi$. Suppose that*

$$\overline{\mathfrak{D}} \cup \mathrm{supp}(F) \cup \mathrm{supp}(div\, u) \subset B_{S_1},$$

$$u|B_S^c \in L^6(B_S^c)^3, \quad \nabla u|B_S^c \in L^2(B_S^c)^9, \quad \pi|B_S^c \in L^2(B_S^c).$$

Then there are coefficients $\beta_1, \beta_2, \beta_3 \in \mathbb{R}$ and functions $\mathfrak{F}_1, \mathfrak{F}_2, \mathfrak{F}_3 \in C^0(B_S^c)$ such that for $j \in \{1, 2, 3\}$, $y \in B_S^c$,

$$u_j(y) = \sum_{k=1}^{3} \beta_k\, \mathcal{Z}_{jk}(y, 0) + \left(\int_{\partial\mathfrak{D}} u \cdot n^{(\mathfrak{D})}\, do_z + \int_{B_{S_1}} div\, u\, dz\right) E_{4j}(y) + \mathfrak{F}_j(y),$$

$$(6.64)$$

and

$$|\mathfrak{F}_j(y)| \leq \mathfrak{C}(S, S_1) \big(\|F\|_1 + \|div\, u\|_1 + \|\nabla u\,|\, \partial\mathfrak{D}\|_1 + \|\pi|\partial\mathfrak{D}\|_1 \tag{6.65}$$
$$+ C(\mathfrak{D}, p) \|u|\partial\mathfrak{D}\|_{2-1/p,p} \big) \big(|y| s_\tau(y) \big)^{-3/2},$$

where $C(\mathfrak{D}, p) > 0$ depends only on $\mathfrak{D}$ and p. (Note that $|E_{4j}(y)| \leq \mathfrak{C}|y|^{-2}$ and $|y|^{-2} \leq \mathfrak{C}(S) \big(|y| s_\tau(y) \big)^{-1}$ for $y \in B_S^c$; see Lemma 3.8.)

Proof Take $j \in \{1, 2, 3\}$, $y \in B_S^c$. Observe that

$$|y - \vartheta z| \geq |y| - S_1 \geq (1 - S_1/S)|y| > 0 \quad \text{for } z \in B_{S_1},\ \vartheta \in [0, 1]. \tag{6.66}$$

In view of Lemma 4.3, we may conclude that the term $\mathcal{Z}_{jk}(y, \vartheta z)$ is continuously differentiable with respect to $\vartheta \in [0, 1]$, for any $z \in B_{S_1}$ and $k \in \{1, 2, 3\}$, with obvious derivatives. Therefore we may define

$$\mathfrak{F}_j(y)$$

$$:= \int_{B_{S_1}} \left(\sum_{k,s=1}^{3} \int_0^1 \partial x_s \mathcal{Z}_{jk}(y, x)_{|x = \vartheta z}\, d\vartheta\, z_s\, F_k(z) \right.$$

$$\left. - \sum_{s=1}^{3} \int_0^1 \partial_s E_{4j}(y - \vartheta z)\, d\vartheta\, z_s\, div\, u(z) \right) dz$$

$$+ \int_{\partial\mathfrak{D}} \left(\sum_{k,s=1}^{3} \int_0^1 \partial x_s \mathcal{Z}_{jk}(y, x)_{|x = \vartheta z}\, d\vartheta\, z_s \right.$$

$$\sum_{l=1}^{3} \big(-\partial_l u_k(z) + \delta_{kl}\, \pi(z) + \partial_l \mathcal{E}_p(u_k)(z) \big) n_l^{(\mathfrak{D})}(z)$$

$$\left. - \sum_{s=1}^{3} \int_0^1 \partial_s E_{4j}(y - \vartheta z)\, d\vartheta\, z_s\, u_k(z) n_k^{(\mathfrak{D})}(z) \right) do_z$$

$$+ \int_{\mathfrak{D}} \sum_{k=1}^{3} \left(\partial_k E_{4j}(y - z) \mathcal{E}_p(u_k)(z) \right.$$

$$+ \sum_{s=1}^{3} \int_0^1 \partial x_s \mathcal{Z}_{jk}(y, x)_{|x = \vartheta z}\, d\vartheta\, z_s \big((\tau e_1 - \omega \times z) \cdot \nabla \mathcal{E}_p(u_k)(z)$$

$$\left. + \big[\omega \times \big(\mathcal{E}_p(u_s)(z) \big)_{1 \leq s \leq 3} \big]_k - \Delta \mathcal{E}_p(u_k)(z) \big) \right) dz,$$

where the extension operator $\mathcal{E}_p$ was introduced in Lemma 6.8. We further set

$$\beta_k := \int_{B_{S_1}} F_k(z)\, dz + \int_{\partial\mathfrak{D}} \sum_{l=1}^{3} \left(-\partial_l u_k(z) + \delta_{kl}\,\pi(z) + \partial_l \mathcal{E}_p(u_k)(z) \right) n_l^{(\mathfrak{D})}(z)\, do_z$$

$$+ \int_{\mathfrak{D}} \Big(\left(\tau e_1 - \omega \times z \right) \cdot \nabla\mathcal{E}_p(u_k)(z)$$

$$+ \big[\omega \times \left(\mathcal{E}_p(u_s)(z) \right)_{1 \le s \le 3} \big]_k - \Delta\mathcal{E}_p(u_k)(z) \Big)\, dz.$$

Then, referring to (6.44), (6.43), (6.54) and (6.55), we obtain (6.64). By (6.66), the choice of $\mathcal{E}_p$ in Lemma 6.8, and (4.33), we further find

$$|\mathfrak{F}_j(y)| \le \mathfrak{C}(S, S_1) \left(|y|\, s_\tau(y) \right)^{-3/2} \Big(\|F\|_1 + \|\nabla u\,|\,\partial\mathfrak{D}\|_1 + \|\pi|\partial\mathfrak{D}\|_1 \tag{6.67}$$

$$+ \sum_{k=1}^{3} \left(\|\nabla\mathcal{E}_p(u_k)\,|\,\partial\mathfrak{D}\|_1 + \|\mathcal{E}_p(u_k)\|_{2,1} \right) \Big)$$

$$+ \mathfrak{C}(S, S_1) |y|^{-3} \Big(\|\mathrm{div}\, u\|_1 + \|u|\partial\mathfrak{D}\|_1 + \sum_{k=1}^{3} \|\mathcal{E}_p(u_k)\|_1 \Big).$$

Inequality (6.67), Lemma 3.8 and (6.60) imply (6.64). ∎

6.3 Representation Formula for the Navier–Stokes System

Finally we use Eq. (6.44) in order to obtain a representation formula for weak solutions of the stationary Navier–Stokes system with Oseen and rotational terms.

Theorem 6.9 Let $u \in W^{1,1}_{loc}(\overline{\mathfrak{D}}^c)^3 \cap L^6(\overline{\mathfrak{D}})^3$, $u_{|\partial\mathfrak{D}} \in W^{2-1/p,p}(\partial\mathfrak{D})$ with $\nabla u \in L^2(\overline{\mathfrak{D}})^9$. Let $\pi \in L^2(\overline{\mathfrak{D}})$, $p \in (1,\infty)$, $q \in (1,2)$, $f : \overline{\mathfrak{D}}^c \mapsto \mathbb{R}^3$ be a function with $f|\mathfrak{D}_T \in L^p(\mathfrak{D}_T)^3$ for $T \in (0,\infty)$ with $\overline{\mathfrak{D}} \subset B_T$, $f|B_S^c \in L^q(B_S^c)^3$ for some $S \in (0,\infty)$ with $\overline{\mathfrak{D}} \subset B_S$.

Suppose that the pair (u, π) is a weak solution of the Navier-Stokes system with Oseen and rotational terms, and with the right–hand side f, that is,

$$\int_{\mathfrak{D}^c} \Big(\nabla u \cdot \nabla\varphi + \left(\tau (u \cdot \nabla)u + \tau\,\partial_1 u - (\omega \times z) \cdot \nabla u + \omega \times u \right) \cdot \varphi + \pi\,\mathrm{div}\,\varphi \Big)\, dz$$

$$= \int_{\mathfrak{D}^c} f \cdot \varphi\, dz \quad \text{for } \varphi \in C_0^\infty(\overline{\mathfrak{D}}^c)^3, \quad \mathrm{div}\, u = 0.$$

Then

$$u_j(y) = \Re_j\big(f - \tau(u \cdot \nabla)u \big)(y) + \mathfrak{B}_j(u, \pi)(y) \tag{6.68}$$

for $j \in \{1, 2, 3\}$, *a.e.* $y \in \overline{\mathfrak{D}}^c$, *where* $\mathfrak{B}_j(u, \pi)$ *was defined in (6.43).*

Proof Since $u \in L^6(\overline{\mathfrak{D}})^3$ and $\nabla u \in L^2(\overline{\mathfrak{D}})^9$, Hölder's inequality yields $\tau(u \cdot \nabla)u \in L^{3/2}(\overline{\mathfrak{D}}^c)^3$. It further follows that the term $\tau \partial_1 u(z) - (\omega \times z) \cdot \nabla u(z) + \omega \times u(z)$, considered as a function of $z \in \mathfrak{D}_T$, belongs to $L^2(\mathfrak{D}_T)^3$ for any $T \in (0, \infty)$ with $\mathfrak{D} \subset B_T$. Therefore, putting

$$F(z) := f(z) - \tau\big(u(z) \cdot \nabla\big)u(z) - \tau \partial_1 u(z) + (\omega \times z) \cdot \nabla u(z) - \omega \times u(z)$$

for $z \in \overline{\mathfrak{D}}^c$, we see that $F|\mathfrak{D}_T \in L^{\min\{p,3/2\}}(\mathfrak{D}_T)^3$, for T as above. Thus, considering the pair (u, π) as a weak solution (in the sense of [27, (IV.1.3)]) of the Stokes system with the right–hand side F, we may refer to [27, Theorem IV.4.1] (interior regularity for the Stokes system), to obtain that

$$u|\mathfrak{D}_T \in W^{2,\min\{p,3/2\}}_{\text{loc}}(\mathfrak{D}_T)^3, \quad \pi|\mathfrak{D}_T \in W^{1,\min\{p,3/2\}}_{\text{loc}}(\mathfrak{D}_T) \quad (T \text{ as above}),$$
$$-\Delta u + \nabla \pi = F, \quad \text{hence} \quad L(u) + \nabla \pi = f - \tau(u \cdot \nabla)u.$$

As $\tau(u \cdot \nabla)u \in L^{3/2}(\overline{\mathfrak{D}}^c)^3$, we now conclude that $L(u) + \nabla \pi | \mathfrak{D}_T \in L^{\min\{p,3/2\}}(\mathfrak{D}_T)^3$ for T as above, so $(u, \pi) \in \mathfrak{M}_{\min\{p,3/2\}}$. The preceding observations mean that the assumptions of Theorem 6.6 are satisfied with p, p_1 replaced by $\min\{p, 3/2\}$ and q, respectively, and with $p_2 = 3/2$. Thus equation (6.68) follows from Theorem 6.6. ∎

6.4 Asymptotic Profile of the Gradient of the Velocity Field

Theorem 6.10 *Let* $p \in (1, \infty)$, $(u, \pi) \in \mathfrak{M}_p$, S, $S_1 \in (0, \infty)$ *with* $S_1 < S$. *Put* $f := L(u) + \nabla \pi$, *and suppose that*

$$\overline{\mathfrak{D}} \cup \text{supp}(f) \cup \text{supp}(div\, u) \subset B_{S_1},$$
$$u|B_S^c \in L^6(B_S^c)^3, \quad \nabla u|B_S^c \in L^2(B_S^c)^9, \quad \pi|B_S^c \in L^2(B_s^c).$$

Then there are coefficients $\beta_1, \beta_2, \beta_3 \in \mathbb{R}$ *and functions* $\mathfrak{F}_1, \mathfrak{F}_2, \mathfrak{F}_3 \in C^1(\overline{B_{S_1}}^c)$ *such that for* $j \in \{1, 2, 3\}$, $\alpha \in \mathbb{N}_0^3$ *with* $|\alpha| \le 1$, $y \in \overline{B_{S_1}}^c$,

$$\partial^\alpha u_j(y) \tag{6.69}$$
$$= \sum_{k=1}^3 \beta_k \cdot \partial_y^\alpha \mathcal{Z}_{jk}(y, 0) + \left(\int_{\partial \mathfrak{D}} u \cdot n^{(\mathfrak{D})}\, do_z + \int_{B_{S_1}} div\, u\, dz \right) \cdot \partial^\alpha E_{4j}(y) + \partial^\alpha \mathfrak{F}_j(y)$$

and if $y \in B_S^c$,

$$|\partial^\alpha \mathfrak{F}_j(y)| \leq \mathfrak{C}(S, S_1) \cdot \left(\|f\|_1 + \|div\,u\|_1 + \|\nabla u\,|\,\partial\mathfrak{D}\|_1 + \|\pi|\partial\mathfrak{D}\|_1 + \|u|\partial\mathfrak{D}\|_1 \right) \tag{6.70}$$

$$\cdot \left(|y| \cdot s_\tau(y) \right)^{-3/2 - |\alpha|/2}.$$

Remark 6.1 The new feature presented by Theorem 6.10 is that the case $|\alpha| = 1$ is admitted.

Proof of Theorem 6.10: Take $j \in \{1, 2, 3\}$, $y \in \overline{B_S}^c$. Observe that

$$|y - \vartheta \cdot z| \geq |y| - S_1 \geq (1 - S_1/S) \cdot |y| > 0 \quad \text{for } z \in B_{S_1}, \ \vartheta \in [0, 1]; \tag{6.71}$$

compare (6.31). Note that (6.71) holds in particular if $z \in \overline{\mathfrak{D}}$. We put

$$\beta_k := \int_{B_{S_1}} f_k(z)\, dz$$

$$+ \int_{\partial\mathfrak{D}} \sum_{l=1}^{3} \left(-\partial_l u_k(z) + \delta_{kl} \cdot \pi(z) + u_k(z) \cdot (\tau \cdot e_1 - \omega \times z)_l \right) \cdot n_l^{(\mathfrak{D})}(z)\, do_z$$

for $1 \leq k \leq 3$,

$$\mathfrak{F}_j(y) := \int_{B_{S_1}} \left(\sum_{k=1}^{3} \left[\left(\mathcal{Z}_{jk}(y, z) - \mathcal{Z}_{jk}(y, 0) \right) \cdot f_k(z) \right] \right. \tag{6.72}$$

$$\left. + \left(E_{4j}(y - z) - E_{4j}(y) \right) \cdot \operatorname{div} u(z) \right) dz$$

$$+ \int_{\partial\mathfrak{D}} \sum_{k=1}^{3} \left(\left(\mathcal{Z}_{jk}(y, z) - \mathcal{Z}_{jk}(y, 0) \right) \right.$$

$$\cdot \sum_{l=1}^{3} \left(-\partial_l u_k(z) + \delta_{kl} \cdot \pi(z) + u_k(z) \cdot (\tau \cdot e_1 - \omega \times z)_l \right) \cdot n_l^{(\mathfrak{D})}(z)$$

$$\left. + \left(E_{4j}(y - z) - E_{4j}(y) \right) \cdot u_k(z) \cdot n_k^{(\mathfrak{D})}(z) \right) do_z$$

$$+ \int_{\partial\mathfrak{D}} \sum_{k,l=1}^{3} \partial_{z_l} \mathcal{Z}_{jk}(y, z) \cdot u_k(z) \cdot n_l^{(\mathfrak{D})}(z)\, dz$$

for $y \in \overline{B_{S_1}}^c$, $1 \leq j \leq 3$. Then, by Lemmas 6.6 and 6.7, we may conclude that $\mathfrak{F}_j \in C^1(\overline{B_{S_1}}^c)$ and the derivative $\partial_{y_m} \mathfrak{F}_j(y)$ equals the right-hand side of (6.72), but with the differential operator ∂_{y_m} acting on $\mathcal{Z}_{jk}(y, z) - \mathcal{Z}_{jk}(y, 0)$, E_{4j} $(y - z) - E_{4j}(y)$ and $\partial_{z_l} \mathcal{Z}_{jk}(y, z)$ ($z \in B_{S_1}$ or $z \in \partial\mathfrak{D}$, $y \in \overline{B_{S_1}}^c$, $1 \leq k, l, m \leq 3$). Obviously Eq. (6.69) holds. Now we recall that $\mathcal{Z}_{jk} \in C^2\left((\mathbb{R}^3 \times \mathbb{R}^3) \backslash \{(x, x) : x \in \mathbb{R}^3\} \right)$ (Lemma 3.10). Thus, in view of (6.71), we find for $y \in B_S^c$, $z \in B_{S_1}$, $\alpha \in \mathbb{N}_0^3$

with $|\alpha| \leq 1$, $1 \leq k \leq 3$ that

$$|\partial_y^\alpha \mathcal{Z}_{jk}(y,z) - \partial_y^\alpha \mathcal{Z}_{jk}(y,0)| = \left| \sum_{s=1}^3 \int_0^1 \partial_y^\alpha \partial_{x_s} \mathcal{Z}_{jk}(y,x)_{|x=\vartheta \cdot z} \cdot z_s \, d\vartheta \right| \quad (6.73)$$

$$\leq \mathfrak{C}(S_1, S) \cdot \left(|y| \cdot s_\tau(y) \right)^{-3/2 - |\alpha|/2},$$

where the last inequality follows from Theorem 4.4; note that $\vartheta \cdot z \in B_{S_1}$ for $z \in B_{S_1}$, $\vartheta \in [0, 1]$. Similarly, since

$$|\partial^\beta E_{4j}(x)| \leq \mathfrak{C} \cdot |x|^{-2-|\beta|} \quad \text{for } x \in \mathbb{R}^3 \backslash \{0\}, \ \beta \in \mathbb{N}_0^3 \text{ with } |\beta| \leq 2,$$

and because of (6.71), we have

$$|\partial_y^\alpha E_{4j}(y-z) - \partial_y^\alpha E_{4j}(y)| = \left| \sum_{s=1}^3 \int_0^1 \partial_x^{\alpha + e_s} E_{4j}(x)_{|x=y-\vartheta \cdot z} \cdot z_s \, d\vartheta \right| \quad (6.74)$$

$$\leq \mathfrak{C} \cdot \int_0^1 |y - \vartheta \cdot z|^{-3-|\alpha|} \, d\vartheta \leq \mathfrak{C}(S_1, S) \cdot |y|^{-3-|\alpha|} \leq \mathfrak{C}(S_1, S) \cdot \left(|y| \cdot s_\tau(y) \right)^{-3/2-|\alpha|/2}$$

for $z \in B_{S_1}$, $\alpha \in \mathbb{N}_0^3$ with $|\alpha| \leq 1$, where the last inequality of (6.74) follows from Lemma 3.8. Further note that by Theorem 4.4,

$$|\partial_y^\alpha \partial_{z_l} \mathcal{Z}_{jk}(y,z)| \leq \mathfrak{C}(S_1, S) \cdot \left(|y| \cdot s_\tau(y) \right)^{-3/2-|\alpha|/2} \quad (6.75)$$

for $z \in \partial \mathfrak{D}$, $\alpha \in \mathbb{N}_0^3$ with $|\alpha| \leq 1$, $1 \leq k, l \leq 3$. Inequalities (6.73)–(6.75) together with (6.72) yield (6.70). $\blacksquare$

6.5 Decay Estimates of the Second Derivatives of the Velocity

Due to the integral $\int_{\partial \mathfrak{D}} \partial_{z_l} \mathcal{Z}_{jk}(y,z) \cdot u_k(z) \cdot n_l^{(\mathfrak{D})}(z) \, dz$ appearing in the definition of $\mathfrak{B}_j(y)$, the second derivatives of $\mathfrak{B}_j(y)$ cannot be evaluated directly because we do not have estimate of the third-order derivatives of $\mathcal{Z}_{jk}$. But the differential operator ∂_{z_l} acting on $\mathcal{Z}_{jk}$ in the above integral may be moved away by a partial integration. We recall the corresponding result as from previous part

Lemma 6.10 (Lemma 6.8) *Let* $p \in (1, \infty)$, $(u, \pi) \in \mathfrak{M}_p$, $j \in \{1, 2, 3\}$. *Define*

$$\mathfrak{U}_j(y) := \mathfrak{U}_j(u)(y) := \int_{\partial \mathfrak{D}} \sum_{k,l=1}^3 \partial_{z_l} \mathcal{Z}_{jk}(y,z) \cdot u_k(z) \cdot n_l^{(\mathfrak{D})}(z) \, do_z$$

for $y \in \overline{\mathfrak{D}}^c$. *Let* $\mathcal{E}_p : W^{2-1/p,p}(\partial \mathfrak{D}) \mapsto W^{2,p}(\mathfrak{D})$ *denote a continuous extension operator ([51]). Then, for* $y \in \overline{\mathfrak{D}}^c$,

$$
\mathfrak{U}_j(y) = \int_{\mathfrak{D}} \sum_{k=1}^{3} \Big[\partial_k E_{4j}(y-z) \cdot \mathcal{E}_p(u_k)(z) + \mathcal{Z}_{jk}(y,z) \cdot \big((\tau \cdot e_1 - \omega \times z) \big.
$$

$$
\cdot \nabla \mathcal{E}_p(u_k)(z) + \big[\omega \times \big(\mathcal{E}_p(u_s)(z) \big)_{1 \le s \le 3} \big]_k - \Delta \mathcal{E}_p(u_k)(z) \Big) \Big] dz
$$

$$
+ \int_{\partial \mathfrak{D}} \sum_{k,l=1}^{3} \mathcal{Z}_{jk}(y,z) \cdot \big((-\tau \cdot e_1 + \omega \times z)_l \cdot u_k(z) + \partial_l \mathcal{E}_p(u_k)(z) \big) \cdot n_l^{(\mathfrak{D})}(z) \, do_z.
$$

Writing $\mathfrak{B}_j(y) = \big(\mathfrak{B}_j(y) - \mathfrak{U}_j(y) \big) + \mathfrak{U}_j(y)$, we see that we have transformed $\mathfrak{B}_j(y)$ in such a way that no derivative acts on $\mathcal{Z}_{jk}$ any more. But on the other hand, the term $\|u|\partial \mathfrak{D}\|_{2-1/p,\,p}$ appears in the upper bound of the terms involving $\mathcal{E}_p(u)$. In fact, we obtain

Lemma 6.11 *Let* $p \in (1, \infty)$, $(u, \pi) \in \mathfrak{M}_p$, $S_1, S \in (0, \infty)$ *with* $\overline{\mathfrak{D}} \subset B_{S_1}$ *and* $S_1 < S$. *Then*

$$
|\partial^\alpha \mathfrak{B}_j(y)|
$$
$$
\le \mathfrak{C}(S_1, S) \cdot \big(\|\nabla u | \partial \mathfrak{D}\|_1 + \|\pi|\partial \mathfrak{D}\|_1 + C_p \cdot \|u|\partial \mathfrak{D}\|_{2-1/p,p} \big) \cdot \big(|y| \cdot s_\tau(y) \big)^{-1-|\alpha|/2}
$$

for $y \in B_S^c$, $1 \le j \le 3$, $\alpha \in \mathbb{N}_0^3$ *with* $|\alpha| \le 2$, *where* C_p *is a constant with* $\|\mathcal{E}_p(v)\|_{2,p} \le C_p \cdot \|v|\partial \mathfrak{D}\|_{2-1/p,\,p}$ *for* $v \in W^{2-1/p,\,p}(\partial \Omega)$. *This constant may, of course, be chosen in such a way that it depends only on* $\mathfrak{D}$ *and* p.

Proof Use Lemmas 6.6 and 6.7 (with r, T replaced by S_1, S, respectively), and note that $|y|^{-1} \le \mathfrak{C}(S) \cdot s_\tau(y)^{-1}$ for $y \in B_S^c$ by Lemma 3.8. ∎

Exploiting the representation formula (6.37), we may now estimate the second derivatives of u, under the assumption that f and div u have compact support:

Theorem 6.11 *Let* $p \in (1, \infty)$, $(u, \pi) \in \mathfrak{M}_p$. *Put* $f := L(u) + \nabla \pi$. *Suppose there are numbers* $S_1, S \in (0, \infty)$, *with* $S_1 < S$, $\overline{\mathfrak{D}} \cup \text{supp}(f) \cup \text{supp}(\text{div } u) \subset B_{S_1}$,

$$
u|B_S^c \in L^6(B_S^c)^3, \quad \nabla u|B_S^c \in L^2(B_S^c)^9, \quad \pi|B_S^c \in L^2(B_S^c).
$$

Let $j \in \{1, 2, 3\}$, $y \in B_S^c$, $\alpha \in \mathbb{N}_0^3$ *with* $|\alpha| \le 2$. *Then*

$$
|\partial^\alpha u_j(y)| \le \mathfrak{C}(S, S_1) \cdot \big(\|f|B_{S_1}\|_1 + \|\text{div } u\|_1 + \|\nabla u | \partial \mathfrak{D}\|_1 + \|\pi|\partial \mathfrak{D}\|_1
$$
$$
+ C_p \cdot \|u|\partial \mathfrak{D}\|_{2-1/p,p} \big) \cdot \big(|y| \cdot s_\tau(y) \big)^{-1-|\alpha|/2},
$$

where C_p *was introduced in Lemma 6.11.*

Proof Starting from (6.37), we estimate the boundary potential $\partial^\alpha \mathfrak{B}_j(y)$ by applying Lemma 6.11 and the volume potentials $\partial^\alpha \mathfrak{R}_j(f)(y)$ and $\partial^\alpha \mathfrak{S}_j(\operatorname{div} u)(y)$ by referring to Lemma 6.7 (with R, T replaced by S_1, S, respectively). In the estimate of $\partial^\alpha \mathfrak{S}_j(\operatorname{div} u)(y)$, we further take account of the inequality $|y|^{-1} \le \mathfrak{C}(S) \cdot s_\tau(y)^{-1}$ (Lemma 3.8). ∎

Chapter 7
Leray Solution

7.1 Introduction

We are interested in a linearized version of (2.1) that reads as follows:

$$\mathcal{L}(u) + \nabla \pi = f, \quad \operatorname{div} u = 0 \quad \text{in } \mathbb{R}^3 \backslash \overline{\mathfrak{D}}, \tag{7.1}$$

It is well known [29] that for data of arbitrary size, both problems (2.1) and (7.1) admit a "Leray solution" characterized by the relations

$$u \in L^6(\mathbb{R}^3 \backslash \overline{\mathfrak{D}})^3, \ \nabla u \in L^2(\mathbb{R}^3 \backslash \overline{\mathfrak{D}})^9, \ \pi \in L^2_{\text{loc}}(\mathbb{R}^3 \backslash \overline{\mathfrak{D}}). \tag{7.2}$$

Galdi, Kyed [30] showed that if the right-hand side f in the nonlinear problem (2.1) has bounded support, then the velocity part u of such a solution (u, π) decays as follows:

$$|\partial^\alpha u(x)| = O\Big[\big(|x| \cdot s_\tau(x)\big)^{-1-|\alpha|/2}\Big] \quad (|x| \to \infty), \tag{7.3}$$

where $\alpha \in \mathbb{N}_0^3$ with $|\alpha| := \alpha_1 + \alpha_2 + \alpha_3 \leq 1$ (decay of u and ∇u). The term $s_\tau(x)$ in (7.3) is defined by

$$s_\tau(x) := 1 + \tau(|x| - x_1) \quad (x \in \mathbb{R}^3). \tag{7.4}$$

Its presence in (7.3) may be considered as a mathematical manifestation of the wake extending downstream behind the rigid body. In the previous section we considered the condition

$$u_{|B_S^c} \in L^6(B_S^c)^3, \ \nabla u_{|B_S^c} \in L^2(B_S^c)^9, \ \pi_{|B_S^c} \in L^2(B_S^c) \tag{7.5}$$

© Atlantis Press and the author(s) 2016

Š. Nečasová and S. Kračmar, *Navier–Stokes Flow Around a Rotating Obstacle*,
Atlantis Briefs in Differential Equations 3, DOI 10.2991/978-94-6239-231-1_7

for some $S > 0$ with $\overline{\mathfrak{D}} \subset B_S$ as well as some additional regularity assumptions, which require in particular that $\pi \in W^{1,p}_{\mathrm{loc}}(\mathbb{R}^3 \backslash \overline{\mathfrak{D}}^c)$ and $\pi_{|B_T \backslash \overline{\mathfrak{D}}} \in L^p(B_T \backslash \overline{\mathfrak{D}})$ for some $p \in (1, \infty)$ and for any $T \in (0, \infty)$ with $\overline{\mathfrak{D}} \subset B_T$. In this section, we will show that the representation formula and decay property remain valid when the condition $\pi_{|B_S^c} \in L^2(B_S^c)$ in (7.5) is dropped.

7.2 Auxiliary Results

The key auxiliary result of this section is

Theorem 7.1 *Let* $q \in (1, 2)$, $p \in (1, \infty)$, $S > 0$, $f \in L^q(\overline{B_S}^c)^3 + L^{3/2}(\overline{B_S}^c)^3$, $u \in L^6(\overline{B_S}^c)^3 \cap W^{2,p}_{\mathrm{loc}}(\overline{B_S}^c)^3$, $\pi \in W^{1,p}_{\mathrm{loc}}(\overline{B_S}^c)$ *with* $L(u) + \nabla \pi = f$, $\mathrm{div}\, u = 0$.
Then there is $c \in \mathbb{R}$ *with* $\pi + c_{|B_{2S}^c} \in L^{3q/(3-q)}(B_{2S}^c) + L^3(B_{2S}^c)$.

Proof We use the approach from the proof of [30, Theorem 4.4]. By the cut-off procedure from that proof and since $W^{1,q}_{\mathrm{loc}}(\mathbb{R}^3) \subset L^{3/2}_{\mathrm{loc}}(\mathbb{R}^3)$, we see there are functions $F \in L^q(\mathbb{R}^3)^3 + L^{3/2}(\mathbb{R}^3)^3$, $U \in L^6(\mathbb{R}^3)^3 \cap W^{2,p}_{\mathrm{loc}}(\mathbb{R}^3)^3$, $\Pi \in W^{1,p}_{\mathrm{loc}}(\mathbb{R}^3)$ such that

$$L(U) + \nabla \Pi = F, \quad \mathrm{div}\, U = 0, \quad U(x) = u(x) + \beta |x|^{-3} x \text{ for } x \in B_{2S}^c,$$

$$\Pi_{|B_{2S}^c} = \pi_{|B_{2S}^c}$$

with some constant $\beta \in \mathbb{R}$. Note that the argument from that proof strongly simplifies in the present situation because we consider the linear problem (7.1) instead of the nonlinear one (2.1). By the assumptions on F, there are $F^{(1)} \in L^q(\mathbb{R}^3)^3$, $F^{(2)} \in L^{3/2}(\mathbb{R}^3)^3$ such that $F = F^{(1)} + F^{(2)}$. But according to [17], there are pairs of functions $(U^{(1)}, \Pi^{(1)}) \in X_q$, $(U^{(2)}, \Pi^{(2)}) \in X_{3/2}$ such that

$$L(U^{(\kappa)}) + \nabla \Pi^{(\kappa)} = F^{(\kappa)}, \quad \mathrm{div}\, U^{(\kappa)} = 0 \text{ for } \kappa \in \{1, 2\}.$$

This means in particular that $U^{(1)} \in W^{2,q}_{\mathrm{loc}}(\mathbb{R}^3)^3 \cap L^{2q/(2-q)}(\mathbb{R}^3)^3$, $\Pi^{(1)} \in W^{1,q}_{\mathrm{loc}}(\mathbb{R}^3)$, $U^{(2)} \in L^6(\mathbb{R}^3)^3$ (hence $U - U^{(2)} \in L^6(\mathbb{R}^3)^3$), and $U^{(2)} \in W^{2,3/2}_{\mathrm{loc}}(\mathbb{R}^3)^3$, $\Pi^{(2)} \in W^{1,3/2}_{\mathrm{loc}}(\mathbb{R}^3)$.
   We may thus apply [30, Lemma 4.1] with $s = \min\{p, q, 3/2\}$, $q_1 = 2q/(2 - q)$, $q_2 = 6$, $f = F^{(1)}$, $(v_1, p_1) = (U^{(1)}, \Pi^{(1)})$, $(v_2, p_2) = (U - U^{(2)}, \Pi - \Pi^{(2)})$. It follows that $U^{(1)} = U - U^{(2)}$, $\Pi^{(1)} = \Pi - \Pi^{(2)} + c$ for some $c \in \mathbb{R}$. Hence $\Pi + c \in L^{3q/(3-q)}(\mathbb{R}^3) + L^3(\mathbb{R}^3)$, so that $\pi + c_{|B_{2S}^c} \in L^{3q/(3-q)}(B_{2S}^c) + L^3(B_{2S}^c)$. ∎

Theorem 7.2 *Let* $S \in (0, \infty)$. *Then* $|\mathcal{Z}(y, z)| \leq \mathfrak{C}(S)\left(|z|\, s_\tau(z)\right)^{-1}$ *for* $y \in \overline{B_S}$, $z \in B_{2S}^c$.

Proof This theorem is a special case of Theorem 4.3. ∎

7.3 Representation Formula for the Leray solution

Theorem 7.3 $\sum_{k=1}^{3} \partial_{z_k} \mathcal{Z}_{jk}(y, z) = 0$ for $1 \le j \le 3$, $y, z \in \mathbb{R}^3$ with $y \ne z$.

Proof It follows from Lemma 4.3. ∎

Theorem 7.4 *Let* $p \in (1, \infty)$, $(u, \pi) \in \mathfrak{M}_p$. *Put* $F := \mathcal{L}(u) + \nabla \pi$ *and suppose there are numbers* $q \in (1, 2)$, $S \in (0, \infty)$ *such that*

$$\overline{\mathfrak{D}} \cup \mathrm{supp}(\mathrm{div}\, u) \subset B_S,$$

$$u_{|B_S^c} \in L^6(B_S^c)^3, \quad \nabla u_{|B_S^c} \in L^2(B_S^c)^9, \quad F_{|B_S^c} \in L^q(B_S^c)^3 + L^{3/2}(B_S^c)^3.$$

Let $j \in \{1, 2, 3\}$ *and put*

$$\mathfrak{B}_j(y) := \mathfrak{B}_j(u, \pi)(y) \tag{7.6}$$

$$:= \int_{\partial \mathfrak{D}} \sum_{k=1}^{3} \left[\sum_{l=1}^{3} \Big(\mathcal{Z}_{jk}(y, z)(-\partial_l u_k(z) + \delta_{kl}\pi(z) + u_k(z)(\tau e_1 - \omega \times z)_l) \right.$$

$$\left. + \partial_{z_l} \mathcal{Z}_{jk}(y, z) u_k(z) \Big) n_l^{(\mathfrak{D})}(z) + E_j(y - z) u_k(z) n_k^{(\mathfrak{D})}(z) \right] do_z$$

for $y \in \overline{\mathfrak{D}}^c$. *Then*

$$u_j(y) = \mathfrak{R}_j(F)(y) + \mathfrak{S}_j(\mathrm{div}\, u)(y) + \mathfrak{B}_j(y) \tag{7.7}$$

for a.e. $y \in \overline{\mathfrak{D}}^c$. *If* $p > 3/2$, (7.7) *holds for any* $y \in \overline{\mathfrak{D}}^c$, *without the restriction "a.e.".*

Proof By Theorem 7.1, there is $c \in \mathbb{R}$, $\pi_1 \in L^{3q/(3-q)}(B_{2S}^c)$, $\pi_2 \in L^3(B_{2S}^c)$ such that $\pi_{|B_{2S}^c} = \pi_1 + \pi_2 + c$. From this fact and our assumptions on u it follows that

$$\int_{2S}^{\infty} \int_{\partial B_r} (|u(z)|^6 + |\nabla u(z)|^2 + |\pi_1(z)|^{3q/(3-q)} + |\pi_2(z)|^3) do_z dr < \infty.$$

Thus there is an increasing sequence (R_n) in $(2S, \infty)$ with $R_n \to \infty$ and

$$\int_{\partial B_{R_n}} (|u(z)|^6 + |\nabla u(z)|^2 + |\pi_1(z)|^{3q/(3-q)} + |\pi_2(z)|^3) do_z \le R_n^{-1} \text{ for } n \in \mathbb{N}. \tag{7.8}$$

By our assumptions on F, there are functions $G^{(1)} \in L^q(B_S^c)^3$, $G^{(2)} \in L^{3/2}(B_S^c)^3$ such that $F_{|B_S^c} = G^{(1)} + G^{(2)}$. Thus by Lemmas 6.1, 6.2

$$\int_{\overline{\mathfrak{D}}^c} \sum_{k=1}^{3} |\mathcal{Z}_{jk}(y,z)| \Big(\chi_{(0,S)}(|z|)|F_k(z)| + \chi_{(S,\infty)}(|z|)(|G_k^{(1)}(z)| + |G_k^{(2)}(z)|) \Big) dz \quad (7.9)$$
$$< \infty$$

for a.e. $y \in \overline{\mathfrak{D}}^c$. Moreover, by Lemma 6.3 with $p = q = 2$

$$\int_{\overline{\mathfrak{D}}^c} |E_j(y-z)| \, |\operatorname{div} u(z)| dz < \infty \qquad (7.10)$$

for a.e. $y \in \overline{\mathfrak{D}}^c$. Due to these observations and Theorem 6.5, we see there is a subset N of $\overline{\mathfrak{D}}^c$ with measure zero such that the relations in (7.9), (7.10) hold for $y \in \overline{\mathfrak{D}}^c \setminus N$ and such that (6.53) with R replaced by R_n holds for $n \in \mathbb{N}$ and $y \in \mathfrak{D}_{R_n} \setminus N$. In the case $p > 3/2$, Lemmas 6.1, 6.2 yields that (6.53), is valid for any $y \in \overline{\mathfrak{D}}^c$, and Theorem 6.5 states that (6.53) with R replaced by R_n is true for $n \in \mathbb{N}$ and for any $y \in \overline{\mathfrak{D}}^c$. Moreover, since $(u,\pi) \in \mathfrak{M}_p$ and in view of Theorem 6.4, we obtain $\operatorname{div} u_{|\mathfrak{D}_T} \in W^{1,p}(\mathfrak{D}_T)$ for $T \in (0,\infty)$ with $\overline{\mathfrak{D}} \subset B_T$. But $\operatorname{supp}(\operatorname{div} u) \subset B_S$, so $\operatorname{div} u \in W^{1,p}(\overline{\mathfrak{D}}^c)$. Thus, if $p > 3/2$, the Sobolev inequality implies there is $s \in (3,\infty]$ with $\operatorname{div} u \in L^s(\overline{\mathfrak{D}}^c)$, so Lemma 6.2, yields that the restriction "a. e." may be dropped in (7.10) as well.

Take $y \in \overline{\mathfrak{D}}^c$ in the case $p > 3/2$ and $y \in \overline{\mathfrak{D}}^c \setminus N$ otherwise. Let $n \in \mathbb{N}$ with $R_n > |y|$ (hence $y \in \mathfrak{D}_{R_n}$). Then, by (6.53) with R replaced by R_n and π by $\pi - c$, we get

$$u_j(y) = \mathfrak{R}_j(F_{|\mathfrak{D}_{R_n}})(y) + \mathfrak{S}_j(\operatorname{div} u_{|\mathfrak{D}_{R_n}})(y) + U_{in,j}(y) + \mathfrak{B}_j(y) \qquad (7.11)$$

with

$$U_{in,j}(y) := \int_{\partial B_{R_n}} \sum_{k=1}^{3} \Big[\sum_{l=1}^{3} \Big(\mathcal{Z}_{jk}(y,z)(\partial_l u_k(z) - $$
$$- \delta_{kl}(\pi - c)(z) - \tau \delta_{1l} u_k(z)) - \partial_{z_l} \mathcal{Z}_{jk}(y,z)(u_k(z)) \Big) \frac{z_l}{R_n} - E_j(y-z)u_k(z)\frac{z_k}{R_n} \Big] do_z,$$

where we used the relation $\sum_{l=1}^{3}(\omega \times z)_l z_l = 0$ for $z \in \partial B_{R_n}$. Concerning $\mathfrak{B}_j(y)$, we note that

$$\int_{\partial \mathfrak{D}} \sum_{l=1}^{3} \mathcal{Z}_{jk}(y,z) n_k^{(\mathfrak{D})}(z) do_z = 0 \quad \text{for} \quad y \in \mathfrak{D}_R, \ 1 \le j \le 3;$$

see Theorems 4.3 and 7.3. Thus the definition of $\mathfrak{B}_j(y)$ need not be modified even though we replace π by $\pi - c$.

Let $n \in \mathbb{N}$ with $\frac{R_n}{4} \geq |y|$. Observe that

$$|U_{in,j}(y)| \leq \mathfrak{C} \sum_{v=1}^{5} \sum_{k=1}^{3} \mathcal{B}_{v,k}(y)$$

with

$$\mathcal{B}_{1,k}(y) := \left(\int_{\partial B_{R_n}} |\mathcal{Z}_{jk}(y,z)|^{6/5} do_z \right)^{5/6} \|u_{|\partial B_{R_n}}\|_6,$$

$$\mathcal{B}_{2,k}(y) = \left(\int_{\partial B_{R_n}} |\mathcal{Z}_{jk}(y,z)|^2 do_z \right)^{1/2} \|\nabla u_{|\partial B_{R_n}}\|_2,$$

$$\mathcal{B}_{3,k}(y) = \left(\int_{\partial B_{R_n}} |\mathcal{Z}_{jk}(y,z)|^{\frac{3q}{4q-3}} do_z \right)^{\frac{4q-3}{3q}} \|\pi_1{}_{|\partial B_{R_n}}\|_{3q/(3-q)},$$

$$\mathcal{B}_{4,k}(y) = \left(\int_{\partial B_{R_n}} |\mathcal{Z}_{jk}(y,z)|^{3/2} do_z \right)^{2/3} \|\pi_2{}_{|\partial B_{R_n}}\|_3,$$

$$\mathcal{B}_{5,k}(y) = \sum_{l=1}^{3} \left(\int_{\partial B_{R_n}} |\partial z_l \mathcal{Z}_{jk}(y,z)|^{6/5} do_z \right)^{5/6} \|u_{|\partial B_{R_n}}\|_6,$$

$$\mathcal{B}_{6,k}(y) = \left(\int_{\partial B_{R_n}} |y-z|^{-12/5} do_z \right)^{5/6} \|u_{|\partial B_{R_n}}\|_6$$

for $k \in \{1, 2, 3\}$. As in the proof of [5, Theorem 4.6], we get

$$|\mathcal{B}_{1,k}(k)| \leq \mathfrak{C}(|y|)R_n^{-1/3}, \quad |\mathcal{B}_{2,k}(y)| \leq \mathfrak{C}(|y|)R_n^{-1}, \quad |\mathcal{B}_{5,k}(y)| \leq \mathfrak{C}(|y|)R_n^{-5/6},$$
$$|\mathcal{B}_{6,k}(y)| \leq \mathfrak{C}(|y|)R_n^{-1/2}.$$

Now applying Theorem 7.2, Lemma 3.5 and (7.8), we obtain

$$\mathcal{B}_{3,k}(y) \leq \mathfrak{C}(|y|) \left(\int_{\partial B_{R_n}} (|z|s_\tau(z))^{-\frac{3q}{4q-3}} do_z \right)^{\frac{4q-3}{3q}} \|\pi_1|\partial B_{R_n}\|_{3q/(3-q)}$$

$$\leq \mathfrak{C}(|y|)R_n^{-1} \left(\int_{\partial B_{R_n}} |s_\tau(z)|^{-\frac{3q}{4q-3}} do_z \right)^{\frac{4q-3}{3q}} R_n^{-\frac{3-q}{3q}}$$

$$\leq \mathfrak{C}(|y|)R_n^{-1} R_n^{\frac{4q-3}{3q}} R_n^{-\frac{3-q}{3q}} \leq \mathfrak{C}(|y|)R^{\frac{2q-6}{3q}},$$

$$\mathcal{B}_{4,k}(y) \leq \mathcal{C}(|y|) \left(\int_{\partial B_{R_n}} (|z| s_\tau(z))^{-3/2} do_z \right)^{2/3} \|\pi_2|\partial B_{R_n}\|_3$$

$$\leq \mathcal{C}(|y|) R_n^{-1} \left(\int_{\partial R_{n}} |s_\tau(z)|^{-3/2} do_z \right)^{2/3} R_n^{-1/3}$$

$$\leq \mathcal{C}(|y|) R^{-2/3}.$$

Thus we conclude that $U_{in,j}(y) \to 0$ for $n \to \infty$. Turning to $\mathfrak{R}_j(F_{|\mathcal{D}_{R_n}})(y)$ and $\mathfrak{S}_j(\operatorname{div} u_{|\mathcal{D}_{R_n}})(y)$ and applying (7.9), (7.10) and Lebesgue's theorem, it follows that

$$\mathfrak{R}_j(F_{|\mathcal{D}_{R_n}})(y) \to \mathfrak{R}_j(F)|(y), \quad \mathfrak{S}_j(\operatorname{div} u_{|\mathcal{D}_{R_n}})(y) \to \mathfrak{S}_j(\operatorname{div} u)(y)$$

for $n \to \infty$. Theorem 7.4 now follows with (7.11). ∎

7.4 Asymptotic Profile of the Linear Case

We begin by considering the decay of the velocity and its gradient in the linear case.

Theorem 7.5 *Let* $p \in (1, \infty)$, $(u, \pi) \in \mathfrak{M}_p$. *Put* $F := L(u) + \nabla \pi$. *Suppose there are numbers* $S_1, S, \gamma \in (0, \infty)$, $A \in [2, \infty)$, $B \in \mathbb{R}$ *such that* $S_1 < S$,

$$\overline{\mathcal{D}} \cup \operatorname{supp}(\operatorname{div} u) \subset B_{S_1}, \quad u_{|B_S^c} \in L^6(B_S^c)^3, \quad \nabla u_{|B_S^c} \in L^2(B_S^c)^9,$$
$$A + \min\{1, B\} \geq 3, \quad |F(z)| \leq \gamma |z|^{-A} s_\tau(z)^{-B} \text{ for } z \in B_{S_1}^c.$$

Let $i, j \in \{1, 2, 3\}$, $y \in B_S^c$. *Then*

$$|u_j(y)| \leq \mathcal{C}(S, S_1, A, B)(\gamma + \|F_{|\mathcal{D}_{S_1}}\|_1 + \|\operatorname{div} u\|_1 \tag{7.12}$$
$$+ \|u_{|\partial\mathcal{D}}\|_1 + \|\nabla u_{|\partial\mathcal{D}}\|_1 + \|\pi_{|\partial\mathcal{D}}\|_1)$$
$$(|y| s_\tau(y))^{-1} l_{A,B}(y),$$

$$|\partial_i u_j(y)| \leq \mathcal{C}(S, S_1, A, B)(\gamma + \|F_{|\mathcal{D}_{S_1}}\|_1 \tag{7.13}$$
$$+ \|u_{|\partial\mathcal{D}}\|_1 + \|\operatorname{div} u\|_1 + \|\nabla u_{|\partial\mathcal{D}}\|_1 + \|\pi_{|\partial\mathcal{D}}\|_1)$$
$$(|y| s_\tau(y))^{-3/2} s_\tau(y)^{\max(0, 7/2 - A - B)} l_{A,B}(y),$$

where $l_{A,B}(y) := 1$ *if* $A + \min\{1, B\} > 3$ *and* $l_{A,B}(y) := \max(1, \ln|y|)$ *if* $A + \min\{1, B\} = 3$.

Proof Theorem 7.5 may be deduced from Theorems 7.4 and 6.7. ∎

Next we turn to the decay of derivatives of the velocity up to order 2.

Theorem 7.6 *Let* $p \in (1, \infty)$, $(u, \pi) \in \mathfrak{M}_p$. *Put* $F := \mathcal{L}(u) + \nabla\pi$. *Suppose there are numbers* S_1, $S \in (0, \infty)$ *with* $S_1 < S$,

$$\overline{\mathfrak{D}} \cup \operatorname{supp}(F) \cup \operatorname{supp}(\operatorname{div} u) \subset B_{S_1}, \quad u_{|B_S^c} \in L^6(B_S^c)^3, \quad \nabla u_{|B_S^c} \in L^2(B_S^c)^9.$$

Let $\mathcal{E}_p : W^{2-1/p,p}(\partial\mathfrak{D}) \mapsto W^{2,p}(\mathfrak{D})$ *be an extension operator with* $\|\mathcal{E}_p(v)\|_{2,p} \leq C_p \|v\|_{2-1/2,p}$ *for* $v \in W^{2-1/p,p}(\partial\mathfrak{D})$, *for some* $C_p > 0$.
Let $j \in \{1, 2, 3\}$, $y \in B_S^c$, $\alpha \in \mathbb{N}_0^3$ *with* $|\alpha| \leq 2$. *Then*

$$
|\partial^\alpha u_j(y)| \leq \mathfrak{C}(S, S_1) \cdot \Big(\|F_{|\mathfrak{D}_{S_1}}\|_1 + \|\operatorname{div} u\|_1 + \|\nabla u_{|\partial\mathfrak{D}}\|_1
$$
$$
+ \|\pi_{|\partial\mathfrak{D}}\|_1 + C_p \cdot \|u_{|\partial\mathfrak{D}}\|_{2-1/p,p} \Big) \cdot \big(|y| \cdot s_\tau(y) \big)^{-1-|\alpha|/2}. \tag{7.14}
$$

Proof Theorem 7.6 follows from Theorems 7.4 and 6.7.

We shall study an asymptotic profile of the velocity and its gradient of the linear case:

Theorem 7.7 *Let* $p \in (1, \infty)$, $(u, \pi) \in \mathfrak{M}_p$, $S, S_1 \in (0, \infty)$ *with* $S_1 < S$. *Put* $F := \mathcal{L}(u) + \nabla\pi$. *Suppose that*

$$\overline{\mathfrak{D}} \cup \operatorname{supp}(F) \cup \operatorname{supp}(\operatorname{div} u) \subset B_{S_1}, \quad u_{|B_S^c} \in L^6(B_S^c)^3, \quad \nabla u_{|B_S^c} \in L^2(B_S^c)^9.$$

Then there are coefficients $\beta_1, \beta_2, \beta_3 \in \mathbb{R}$ *and functions* $\mathfrak{F}_1, \mathfrak{F}_2, \mathfrak{F}_3 \in C^1(\overline{B_{S_1}}^c)$ *such that for* $j \in \{1, 2, 3\}$, $\alpha \in \mathbb{N}_0^3$ *with* $|\alpha| \leq 1$,

$$
\partial^\alpha u_j(y) = \sum_{k=1}^3 \beta_k \partial_y^\alpha Z_{jk}(y, 0) + \left(\int_{\partial\mathfrak{D}} u \cdot n^{(\mathfrak{D})} do_z + \int_{\mathfrak{D}_{S_1}} \operatorname{div} u \, dz \right) \partial^\alpha E_j(y) + \partial^\alpha \mathfrak{F}_j(y)
$$

for $y \in \overline{B_{S_1}}^c$, *and*

$$
|\partial^\alpha \mathfrak{F}_j(y)| \leq \mathfrak{C}(S, S_1)(\|F\|_1 + \|\operatorname{div} u\|_1 + \|\nabla u_{|\partial\mathfrak{D}}\|_1
$$
$$
+ \|\pi_{|\partial\mathfrak{D}}\|_1 + \|u_{|\partial\mathfrak{D}}\|_1)(|y| s_\tau(y))^{-3/2-|\alpha|/2} \quad \text{for} \quad y \in B_S^c.
$$

Proof Theorem 7.7 may be deduced from Theorem 7.4 in the same way as [9, Theorem 1.1] from [5, Theorem 4.6]. ∎

Remark 7 An explicit formula for β_i, $\mathfrak{F}_i$, $i = 1, \ldots, 3$, is given in [9, page 473].

We may use Theorems 7.5 and 7.6 in order to derive a decay estimate as in (5.1.) for Leray solutions of the linear problem (7.1). This may be done by considering the restriction of such a solution to $B_{S_0}^c$ for some $S_0 > 0$ sufficiently large. The idea is to apply Theorems 7.5 and 7.6 with $\mathfrak{D}$ replaced by B_{S_0}. In this way, the behaviour of the solution in question near the boundary of its domain, and the regularity of that boundary do not matter. A crucial technical result in this respect is the ensuing observation on the interior regularity of generalized solutions to (7.1).

Theorem 7.8 *Let* $U \subset \mathbb{R}^3$ *be open and bounded,* $p \in (1, \infty)$, $f \in L^p_{\text{loc}}(\overline{U}^c)^3$, $u \in W^{1,2}_{\text{loc}}(\overline{U}^c)^3$, $\pi \in L^p_{\text{loc}}(\overline{U}^c)$,

$$\int_{\overline{U}^c} (\nabla u \cdot \nabla \varphi + (\tau \partial_1 u - (\omega \times x) \cdot \nabla u + \omega \times u)\varphi - \pi \text{div}\, \varphi - f\varphi)dx = 0$$

for $\varphi \in C^\infty_0(\overline{U}^c)^3$, $\quad \text{div}\, u = 0$.

$$(7.15)$$

Then $u \in W^{2,\min\{2,p\}}_{\text{loc}}(\overline{U}^c)^3$, $\pi \in W^{1,\min\{2,p\}}_{\text{loc}}(\overline{U}^c)$, *and* $L(u) + \nabla \pi = f$.

Proof This theorem follows from the interior regularity of Stokes flows, as stated in [Theorem IV.4.1]. Details of the argument may be found in the proof of Theorem 6.9. ∎

Now we may establish (5.1) for Leray solutions of (7.1).

Corollary 7.1 *Let* $U \subset \mathbb{R}^3$ *be open and bounded. Let* $p \in (1, \infty)$, $f \in L^p_{\text{loc}}(\overline{U}^c)^3$, $\gamma, S_1 \in (0, \infty)$ *with* $\overline{U} \subset B_{S_1}$, $A \in [2, \infty)$, $B \in \mathbb{R}$ *with* $A + \min\{1, B\} > 3$, $|f(z)| \le \gamma \cdot |z|^{-A} s_\tau(z)^{-B}$ *for* $z \in B^c_{S_1}$.

Let $u \in W^{1,1}_{\text{loc}}(\overline{U}^c)^3$ *with* $u \in L^6(\overline{U}^c)^3$ *and* $\nabla u \in L^2(\overline{U}^c)^9$. *Let* $\pi \in L^2_{\text{loc}}(\overline{U}^c)$ *and suppose that (7.15) holds.*

Choose some $S_0 \in (0, S_1)$ *with* $\overline{U} \subset B_{S_0}$ *and let* $S \in (S_1, \infty)$. *Then, for* $z \in B^c_S$, $1 \le i, j \le 3$, *inequalities (7.12) and (7.13) are valid, but with* $\mathfrak{D}$ *replaced by* $B^c_{S_0}$.

Now, suppose that $\text{supp}(f) \subset B_{S_1}$, *put* $s := s(p) := \min\{2, p\}$, *and let* $\mathcal{E}_s : W^{2-1/s,s}(\partial B_{S_0}) \mapsto W^{2,s}(B_{S_0})$ *be a continuous extension operator. Let* $C_p > 0$ *be a constant with*

$$\|\mathcal{E}_s(v)\|_{2,s} \le C_p \cdot \|v\|_{2-1/s,s} \text{ for } v \in W^{2-1/s,s}(\partial B_{S_0}).$$

Then, for $1 \le j \le 3$, $\alpha \in \mathbb{N}^3_0$ *with* $|\alpha| \le 2$, $x \in B^c_S$, *inequality (7.14) holds with* $\mathfrak{D}$ *replaced by* B_{S_0}, *and with the norm* $\| \cdot \|_{2-1/s,s}$ *in the place of* $\| \cdot \|_{2-1/p,p}$.

Proof Theorem 7.8 yields that $u \in W^{2,\min\{2,p\}}_{\text{loc}}(\overline{U}^c)^3$, $\pi \in W^{1,\min\{2,p\}}_{\text{loc}}(\overline{U}^c)$, and $\mathcal{L}(u) + \nabla \pi = f$. We may conclude in particular that the pair $(u_{|B^c_{S_0}}, \pi_{|B^c_{S_0}})$ belongs to $\mathfrak{M}_{\min\{2,p\}}$, with $\overline{B_{S_0}}^c$ in the place of $\overline{\mathfrak{D}}^c$ as the domain of reference in the definition of $\mathfrak{M}_{\min\{2,p\}}$. Therefore the corollary follows from Theorems 7.5 and 7.6. ∎

7.5 Representation Formula for the Nonlinear Case

In this section, we show that the representation formula Theorem 6.9 pertaining to the velocity part of solutions to the nonlinear problem remains valid even if the pressure does not belong to $L^2(B^c_S)$ for some $S > 0$ with $\overline{\mathfrak{D}} \subset B_S$.

We consider the system of equations

$$-\Delta u + \tau \partial_1 u - (\omega \times x) \cdot \nabla u + \omega \times u + \tau u \cdot \nabla u + \nabla \pi = f, \\ \operatorname{div} u = 0, \qquad (7.16)$$

in the exterior domain $\overline{\mathfrak{D}}^c := \mathbb{R}^3 \backslash \overline{\mathfrak{D}}$, supplemented by a decay condition at infinity, $u(x) \to 0$ for $|x| \to \infty$, and suitable boundary conditions on $\partial \mathfrak{D}$.

Theorem 7.9 *Let* $u \in W^{1,1}_{\mathrm{loc}}(\overline{\mathfrak{D}}^c)^3 \cap L^6(\overline{\mathfrak{D}}^c)^3$ *with* $\nabla u \in L^2(\overline{\mathfrak{D}}^c)^9$.

Let $p \in (1, \infty)$, $q \in (1, 2)$, $f : \overline{\mathfrak{D}}^c \mapsto \mathbb{R}^3$ *a function with* $f_{|\mathfrak{D}_T} \in L^p(\mathfrak{D}_T)^3$ *for* $T \in (0, \infty)$ *with* $\overline{\mathfrak{D}} \subset B_T$, *and* $f_{|B^c_S} \in L^q(B^c_S)^3$ *for some* $S \in (0, \infty)$ *with* $\overline{\mathfrak{D}} \subset B_S$.

Further assume that $u_{|\partial \mathfrak{D}} \in W^{2-1/p,p}(\partial \mathfrak{D})^3$ *and that* $\pi : \overline{\mathfrak{D}}^c \mapsto \mathbb{R}$ *is a function with* $\pi_{|\mathfrak{D}_T} \in L^p(\mathfrak{D}_T)$ *for* T *as above. Suppose that the pair* (u, π) *is a generalized solution of (7.16), that is,*

$$\int_{\overline{\mathfrak{D}}^c} \Big((\nabla u \cdot \nabla \varphi) + (\tau (u \cdot \nabla) u + \tau \partial_1 u - (\omega \times x) \cdot \nabla u + \omega \times u) \cdot \varphi + \pi \operatorname{div} \varphi \Big) dx$$

$$= \int_{\overline{\mathfrak{D}}^c} f \cdot \varphi \, dx \quad \text{for } \varphi \in C^\infty_0(\overline{\mathfrak{D}}^c)^3,$$

$$\operatorname{div} u = 0.$$

Then

$$u_j(y) = \mathfrak{R}_j(f - \tau (u \cdot \nabla) u)(y) + \mathfrak{B}_j(u, \pi)(y) \qquad (7.17)$$

for $j \in \{1, 2, 3\}$ *and for a.e.* $y \in \overline{\mathfrak{D}}^c$, *where* $\mathfrak{B}_j(u, \pi)$ *was defined in (7.6).*

Proof Since $u \in L^6(\overline{\mathfrak{D}}^c)^3$ and $\nabla u \in L^2(\overline{\mathfrak{D}}^c)^9$, we have $(u \cdot \nabla) u \in L^{3/2}(\overline{\mathfrak{D}}^c)^3$, hence $f - \tau \cdot u \cdot \nabla u_{|\mathfrak{D}_T} \in L^{\min\{p,3/2\}}(\mathfrak{D}_T)^3$ for T as in the theorem. By Theorem 7.8 we thus get $u \in W^{2,\min\{p,3/2\}}_{\mathrm{loc}}(\overline{\mathfrak{D}}^c)^3$, $\pi \in W^{1,\min\{p,3/2\}}_{\mathrm{loc}}(\overline{\mathfrak{D}}^c)$, so we may conclude that $(u, \pi) \in \mathfrak{M}_{\min\{p,3/2\}}$. Moreover $f - \tau \cdot u \cdot \nabla u_{|B^c_S} \in L^q(B^c_S)^3 + L^{3/2}(B^c_S)^3$. Thus (7.17) follows from Theorem 7.4. ∎

Chapter 8
Latest Results

We will consider the set of dimensionless equations (see [31])

$$-\Delta u + \tau\, \partial_1 u + \tau\, (u \cdot \nabla)u - (\omega \times x) \cdot \nabla u + \omega \times u + \nabla \pi = f,$$
$$\operatorname{div} u = 0, \tag{8.1}$$

in the exterior domain $\overline{\mathfrak{D}}^c := \mathbb{R}^3 \backslash \mathfrak{D}$ supplemented by a decay condition at infinity,

$$u(x) \to 0 \quad \text{for } |x| \to \infty, \tag{8.2}$$

and suitable boundary conditions on $\partial\mathfrak{D}$.

We are interested in "Leray solutions" of (8.1), (8.2), that is, weak solutions characterized by the conditions $u \in L^6(\overline{\mathfrak{D}}^c)^3 \cap W^{1,1}_{loc}(\overline{\mathfrak{D}}^c)^3$, $\nabla u \in L^2(\overline{\mathfrak{D}}^c)^9$ and $\pi \in L^2_{loc}(\overline{\mathfrak{D}}^c)$.

From [10, 30] it follows that the velocity part u of a Leray solution (u, π) to (8.1), (8.2) decays for $|x| \to \infty$ as expressed by the estimates

$$|u(x)| \le C \left(|x|\, s(x) \right)^{-1}, \quad |\nabla u(x)| \le C \left(|x|\, s(x) \right)^{-3/2} \tag{8.3}$$

for $x \in \mathbb{R}^3$ with $|x|$ sufficiently large, where $s(x) := 1 + |x| - x_1$ $(x \in \mathbb{R}^3)$ and $C > 0$ a constant independent of x. The factor $s(x)$ may be considered as a mathematical manifestation of the wake extending downstream behind a body moving in a viscous fluid.

By Kyed [47] it was shown that

$$\left. \begin{array}{l} u_j(x) = \gamma\, E_{j1}(x) + r_j(x), \\ \partial_l u_j(x) = \gamma\, \partial_l E_{j1}(x) + s_{jl}(x) \end{array} \right\} \quad (x \in \overline{\mathfrak{D}}^c,\ 1 \le j, l \le 3), \tag{8.4}$$

where $E : \mathbb{R}^3 \backslash \{0\} \mapsto \mathbb{R}^4 \times \mathbb{R}^3$ denotes a fundamental solution to the Oseen system

$$-\Delta v + \tau\, \partial_1 v + \nabla\Pi = f, \quad \operatorname{div} v = 0 \quad \text{in } \mathbb{R}^3. \tag{8.5}$$

© Atlantis Press and the author(s) 2016
Š. Nečasová and S. Kračmar, *Navier–Stokes Flow Around a Rotating Obstacle*,
Atlantis Briefs in Differential Equations 3, DOI 10.2991/978-94-6239-231-1_8

The definition of the function E is stated in [27, Sect. 2.1]. As becomes apparent from this definition, the term $E_{j1}(x)$ may be expressed explicitly in terms of elementary functions. The coefficient γ is also given explicitly, its definition involving the Cauchy stress tensor. The remainder terms r and s are characterized by the relations $r \in L^q(\overline{\mathfrak{D}}^c)^3$ for $q \in (4/3, \infty)$, $s \in L^q(\overline{\mathfrak{D}}^c)^3$ for $q \in (1, \infty)$. Since it is known from [27, Sect. 7.3] that $E_{j1}|B_r^c \notin L^q(B_r^c)$ for $r > 0$, $q \in [1, 2]$, and $\partial_l E_{j1}|B_r^c \notin L^q(B_r^c)$ for $r > 0$, $q \in [1, 4/3]$, $j, l \in \{1, 2, 3\}$, the function r decays faster than E_{j1}, and s_{jl} faster than $\partial_l E_{j1}$, in the sense of L^q-integrability. Thus the equations in (8.4) may in fact be considered as asymptotic expansions of u and ∇u, respectively. The theory in [47] is valid under the assumption that u verifies the boundary conditions

$$u(x) = e_1 + (\omega \times x) \quad \text{for } x \in \partial\mathfrak{D} \tag{8.6}$$

and f vanishes. Moreover, reference [47] does not deal with pointwise decay of r and s, but in [48], Kyed indicates that $|r(x)|$ behaves as $O(|x|^{-4/3+\epsilon})$ if $|x| \to \infty$, for some arbitrary but fixed $\epsilon > 0$.

In Theorem 8.1 below we derive a pointwise decay of respectively u and ∇u, which is independent of the boundary conditions. In comparison with [47] and indicated in (8.4) our leading term is less explicit than the term $\gamma E_{j1}(x)$ in (8.4) and instead of the fundamental solution $E_{j1}(x)$ of the stationary Oseen system, we use the time integral of the fundamental solution of the evolutionary Oseen system.

In [12] it was shown that $\mathcal{Z}_{j1}(x, 0) = E_{j1}(x)$ for $x \in \mathbb{R}^3 \setminus \{0\}$, $1 \leq j \leq 3$, and $\lim_{|x|\to\infty} |\partial_x^\alpha \mathcal{Z}_{jk}(x, 0)| = O\big((|x| s(x))^{-3/2-|\alpha|/2}\big)$ for $1 \leq j \leq 3$, $k \in \{2, 3\}$, ([12] Corollary 4.5, Theorem 5.1). Thus, setting

$$\mathfrak{G}_j(x) := \sum_{k=2}^{3} \beta_k \, \mathcal{Z}_{jk}(x, 0) + \mathfrak{F}_j(x) \quad (x \in \overline{B_{S_1}}^c, \ 1 \leq j \leq 3), \tag{8.7}$$

we may deduce from (8.10) that

$$u_j(x) = \beta_1 \, E_{j1}(x) + \left(\int_{\partial\Omega} u \cdot n \, do_x\right) x_j \, (4\pi |x|^3)^{-1} + \mathfrak{G}_j(x), \tag{8.8}$$

for $x \in \overline{B_{S_1}}^c$, $1 \leq j \leq 3$, and

$$\lim_{|x|\to\infty} |\partial^\alpha \mathfrak{G}(x)| = O\big((|x| s(x))^{-3/2-|\alpha|/2} \ln(2 + |x|)\big) \tag{8.9}$$

for $\alpha \in \mathbb{N}_0^3$ with $|\alpha| \leq 1$, (Theorem 5.1, Corollary 8.1). If we compare how the coefficient γ from (8.4) is defined in [47], and the coefficient β_1 from (8.8) in [11] (see Theorem 8.1 below), and if we take account of the boundary condition (8.6) satisfied by u in [47], we see that γ and β_1 coincide.

8.1 Statement of the Main Result

Theorem 8.1 *Let $\mathfrak{D} \subset \mathbb{R}^3$ be open, $p \in (1, \infty)$, $f \in L^p(\mathbb{R}^3)^3$ with supp(f) compact. Let $S_1 \in (0, \infty)$ with $\overline{\mathfrak{D}} \cup \text{supp}(f) \subset B_{S_1}$.*
Let $u \in L^6(\overline{\mathfrak{D}}^c)^3 \cap W_{loc}^{1,1}(\overline{\mathfrak{D}}^c)^3$, $\pi \in L_{loc}^2(\overline{\mathfrak{D}}^c)$ with $\nabla u \in L^2(\overline{\mathfrak{D}}^c)^9$, $\text{div}\, u = 0$
and

$$\int_{\overline{\mathfrak{D}}^c} \Big[\nabla u \cdot \nabla \varphi + \big(\tau\, \partial_1 u + \tau\, (u \cdot \nabla) u - (\omega \times x) \cdot \nabla u + \omega \times u \big) \cdot \varphi - \pi\, \text{div}\, \varphi \Big] dx$$

$$= \int_{\overline{\mathfrak{D}}^c} f \cdot \varphi\, dz \quad \text{for } \varphi \in C_0^\infty(\overline{\mathfrak{D}}^c)^3.$$

(This means the pair (u, π) is a Leray solution to (8.1), (8.2).) Suppose in addition that

$$\mathfrak{D} \text{ is } C^2\text{-bounded}, \quad u|\partial\mathfrak{D} \in W^{2-1/p,\, p}(\partial\mathfrak{D})^3, \quad \pi|B_{S_1}\backslash\overline{\mathfrak{D}} \in L^p(B_{S_1}\backslash\overline{\mathfrak{D}}).$$

Let n denote the outward unit normal to $\mathfrak{D}$, and define

$$\beta_k := \int_{\overline{\mathfrak{D}}^c} f_k(y)\, dy$$

$$+ \int_{\partial\mathfrak{D}} \sum_{l=1}^{3} \big(-\partial_l u_k(y) + \delta_{kl}\, \pi(y) + (\tau\, e_1 - \omega \times y)_l\, u_k(y) - \tau\, (u_l\, u_k)(y) \big)\, n_l(y)\, do_y$$

for $1 \leq k \leq 3$,

$$\mathfrak{F}_j(x) := \int_{\overline{\mathfrak{D}}^c} \Big[\sum_{k=1}^{3} \big(\mathcal{Z}_{jk}(x, y) - \mathcal{Z}_{jk}(x, 0) \big)\, f_k(y) - \tau \cdot \sum_{k,l=1}^{3} \mathcal{Z}_{jk}(x, y)\, (u_l\, \partial_l u_k)(y) \Big]\, dy$$

$$+ \int_{\partial\mathfrak{D}} \sum_{k=1}^{3} \Big[(\mathcal{Z}_{jk}(x, y) - \mathcal{Z}_{jk}(x, 0)) \sum_{l=1}^{3} (-\partial_l u_k(y) + \delta_{kl}\, \pi(y) + (\tau\, e_1 - \omega \times y)_l u_k(y))\, n_l(y)$$

$$+ \big(E_{4j}(x - y) - E_{4j}(x) \big)\, u_k(y)\, n_k(y)$$

$$+ \sum_{l=1}^{3} \big(\partial_{y_l} \mathcal{Z}_{jk}(x, y)\, (u_k\, n_l)(y) + \tau \mathcal{Z}_{jk}(x, 0)\, (u_l\, u_k\, n_l)(y) \big) \Big]\, do_y$$

for $x \in \overline{B_{S_1}}^c$, $1 \leq j \leq 3$. The preceding integrals are absolutely convergent. Moreover $\mathfrak{F} \in C^1(\overline{B_{S_1}}^c)^3$ and equation

$$u_j(x) = \sum_{k=1}^{3} \beta_k\, \mathcal{Z}_{jk}(x, 0) + \Big(\int_{\partial\Omega} u \cdot n\, do_x \Big)\, x_j\, (4\pi\, |x|^3)^{-1} + \mathfrak{F}_j(x)$$

holds. In addition, for any $S \in (S_1, \infty)$, there is a constant $C > 0$ which depends on τ, ϱ, S_1, S, f, u and π, and which is such that

$$|\partial^\alpha \mathfrak{F}(x)| \le C \left(|x| \, s(x) \right)^{-3/2 - |\alpha|/2} \ln(2 + |x|) \quad \text{for } x \in \overline{B_S}^c, \ \alpha \in \mathbb{N}_0^3 \text{ with } |\alpha| \le 1.$$

Proof We will just give main lines of proof. For complete details see [11, 3.1]. The main tool of the proof is the representation formula

$$u_j(y) = \mathcal{R}_j(f - \tau(u \cdot \nabla)u)(y) + \mathcal{B}_j(u, \pi(y)), \tag{8.10}$$

where

$$\mathcal{B}_j(y) := \mathcal{B}_j(u, \pi)(y) \tag{8.11}$$

$$:= \int_{\partial \mathfrak{D}} \sum_{k=1}^3 \left[\sum_{l=1}^3 \left(\mathcal{Z}_{jk}(y, z) \cdot \left(-\partial_l u_k(z) + \delta_{kl} \cdot \pi(z) + u_k(z) \cdot (\tau \cdot e_1 - \omega \times z)_l \right) \right. \right.$$

$$\left. \left. + \partial_{z_l} \mathcal{Z}_{jk}(y, z) \cdot u_k(z) \right) \cdot n_l^{(\mathfrak{D})}(z) \, + \, E_{4j}(y - z) \cdot u_k^{\cdot}(z) \cdot n_k^{(\mathfrak{D})}(z) \right] do_z$$

for $y \in \overline{\mathfrak{D}}^c$, with outer normal $n^{(\mathfrak{D})}$ to $\mathfrak{D}$.

We have to consider the term $\mathfrak{R}_j((u \cdot \nabla)u)(x) = \int_{\overline{\mathfrak{D}}^c} \sum_{k=1}^3 \mathcal{Z}_{jk}(x, y)[(u \cdot \nabla)u_k](y)dy$. Applying the integration by parts, Lemma 6.5, Theorems 4.1, 6.9, 6.1 and 6.10 we get the corresponding estimate. Concerning the leading term for gradient of velocity $\int_{\overline{\mathfrak{D}}^c} \sum_{k=1}^3 \partial_m \mathcal{Z}_{jk}(x, y)[(u \cdot \nabla)u_k](y)dy$ we have first to divide the integral into two parts $\int_{\overline{\mathfrak{D}}^c \setminus B_1(x)}$ and $\int_{B_1(x)}$. Again integrating by parts and using Theorems 6.1, 6.8 and 6.10 we get the leading term for the velocity part.

Theorem 8.2 *Let $\mathfrak{D}$, p, f, S_1, u, π satisfy the assumptions of Theorem 8.1, including (8.10). Let β_1, β_2, β_3 and $\mathfrak{F}$ be defined as in Theorem 8.1. Define the function $\mathfrak{G}$ as*

$$\mathfrak{G}_j(x) := \sum_{k=2}^3 \beta_k \, \mathcal{Z}_{jk}(x, 0) + \mathfrak{F}_j(x) \quad (x \in \overline{B_{S_1}}^c, \ 1 \le j \le 3). \tag{8.12}$$

Then $\mathfrak{G} \in C^1(\overline{B_{S_1}}^c)^3$, the equation

$$u_j(x) = \beta_1 \, E_{j1}(x) + \left(\int_{\partial \Omega} u \cdot n \, do_x \right) x_j \, (4\pi |x|^3)^{-1} + \mathfrak{G}_j(x) \quad (x \in \overline{B_{S_1}}^c, \ 1 \le j \le 3) \tag{8.13}$$

holds, and for any $S \in (S_1, \infty)$, there is a constant $C > 0$ which depends on τ, ϱ, S_1, S, f, u and π, and which is such that

$$|\partial^\alpha \mathfrak{G}(x)| \le C \left(|x| \, s(x) \right)^{-3/2 - |\alpha|/2} \ln(2 + |x|) \quad \text{for } x \in \overline{B_S}^c, \ \alpha \in \mathbb{N}_0^3 \text{ with } |\alpha| \le 1.$$

Corollary 8.1 *Take* $\mathfrak{D}$, p, f, S_1, u, π *as in Theorem 8.1, but without requiring (8.10). (This means that (u, π) is only assumed to be a Leray solution of (8.1), (8.2).) Put $\widetilde{p} := \min\{3/2,\, p\}$.*

Then $u \in W_{loc}^{2,\widetilde{p}}(\overline{\mathfrak{D}}^c)^3$ and $\pi \in W_{loc}^{1,\widetilde{p}}(\overline{\mathfrak{D}}^c)$.

Fix some number $S_0 \in (0, S_1)$ with $\overline{\mathfrak{D}} \cup supp(f) \subset B_{S_0}$, and define β_1, β_2, β_3 and $\mathfrak{F}$ as in Theorem 8.1, but with $\mathfrak{D}$ replaced by B_{S_0} and $n(x)$ by $S_0^{-1}\,x$, for $x \in \partial B_{S_0}$. Moreover, define $\mathfrak{G}$ as in (8.12).

Then all the conclusions of Theorem 8.2 are valid.

Proof Proof is based on the use of Fourier transform of the Oseen resolvent to get Fourier transform of our fundamental solution and applying [36, Lemma 13]. For complete details see [12].

References

1. Adams, A.: *Sobolev Spaces*, Academic Press, New York, 1975.
2. Babenko, K. I., Vasil'ev, M. M.: On the asymptotic behavior of a steady flow of viscous fluid at some distance from an immersed body. J. Appl. Math. Mech. **37** (1973), 651–665.
3. Deuring, P.: Exterior stationary Navier-Stokes flows in 3D with nonzero velocity at infinity: asymptotic behavior of the second derivatives of the velocity. Comm. Partial Diff. Equ. **30** (2005), 987–1020.
4. Deuring, P.: The single-layer potential associated with the time-dependent Oseen system. Proceedings of the 2006 IASME/WSEAS International Conference on Continuum Mechanics. Chalkida, Greece, May 11-13, (2006), 117–125.
5. Deuring, P., Kračmar, S., Nečasová, Š.: On pointwise decay of linearized stationary incompressible viscous flow around rotating and translating bodies. SIAM J. Math. Anal. **43** (2011), 705–738.
6. Deuring, P., Kračmar, S.: Artificial boundary conditions for the Oseen system in 3D exterior domains. Analysis. **20** (2000), 65–90.
7. Deuring, P., Kračmar, S.: Exterior stationary Navier-Stokes flows in 3D with non-zero velocity at infinity: approximation by flows in bounded domains. Math. Nachr. **269-270** (2004), 86–115.
8. Deuring, P., Kračmar, S., Nečasová, Š.: A linearized system describing stationary incompressible viscous flow around rotating and translating bodies: improved decay estimates of the velocity and its gradient. Dynamical Systems, Differential Equations and Applications, Vol. I, Ed. by W. Feng, Z. Feng, M. Grasselli, A. Ibragimov, X. Lu, S. Siegmund and J. Voigt, Discrete Contin. Dyn. Syst., Supplement 2011, 8th AIMS Conference, Dresden, Germany, 351–361.
9. Deuring, P., Kračmar, S., Nečasová, Š.: Linearized stationary incompressible flow around rotating and translating bodies: asymptotic profile of the velocity gradient and decay estimate of the second derivatives of the velocity. J. Differential Equations. **252** (2012), 459–476.
10. Deuring, P., Kračmar, S., Nečasová, Š.: Pointwise decay of stationary rotational viscous incompressible flows with nonzero velocity at infinity. J. Differential Equations. **255** (2013), 1576–1606.
11. Deuring, P., S. Kračmar, Nečasová, Š.: Leading terms of velocity and its gradient of the stationary rotational viscous incompressible flows with nonzero velocity at infinity. arXiv:1511.03916
12. Deuring, P., Kračmar, S., Nečasová, Š.: Asymptotic structure of viscous incompressible flow around a rotating body, with nonvanishing flow field at infinity. arXiv:1511.04378.
13. Dintelmann, E., Geissert, M., Hieber, M.: Strong Lp-solutions to the Navier-Stokes flow past moving obstacles: the case of several obstacles and time dependent velocity. Trans. Amer. Math. Soc. 361 (2009), **2**, 653–669.

© Atlantis Press and the author(s) 2016

Š. Nečasová and S. Kračmar, *Navier–Stokes Flow Around a Rotating Obstacle*,
Atlantis Briefs in Differential Equations 3, DOI 10.2991/978-94-6239-231-1

14. Farwig, R.: A variational approach in weighted Sobolev spaces to the operator $-\Delta + \partial/\partial x_1$ in exterior domains of $\mathbb{R}^3$. Math. Z. **210** (1992), 449–464.

15. Farwig, R.: The stationary exterior 3D-problem of Oseen and Navier-Stokes equations in anisotropically weighted Sobolev spaces. Math. Z. **211** (1992), 409–447.

16. Farwig, R., Hishida, T., Müller, D.: L^q-theory of a singular "winding" integral operator arising from fluid dynamics. Pacific J. Math. **215** (2004), 297–312.

17. Farwig, R.: An L^q-analysis of viscous fluid flow past a rotating obstacle. Tôhoku Math. J. **58** (2005), 129–147.

18. Farwig, R.: Estimates of lower order derivatives of viscous fluid flow past a rotating obstacle. Banach Center Publications Warsaw. **70** (2005), 73–82.

19. Farwig, R., Krbec, M., Nečasová, Š.: A weighted L^q approach to Stokes flow around a rotating body. Ann. Univ. Ferrara, Sez. VII. **54** (2008), 61–84.

20. Farwig, R., Krbec, M., Nečasová, Š.: A eighted L^q-approach to Oseen flow around a rotating body. Math. Methods in the Appl. Sci. **31** 5, (2008), 551–574.

21. Finn, R.: Estimates at infinity for stationary solutions of the Navier-Stokes equations. Bull. Math. de la Soc. Sci. Math. Phys. de la R. P. Roumaine **3** (1959), 387–418.

22. Finn, R.: On the exterior stationary problem for the Navier-Stokes equations, and associated perturbation problems. Arch. Rational Mech. Anal. **19** (1965), 363–406.

23. Farwig, R., Hishida, T.: Stationary Navier-Stokes flow around a rotating obstacles. Funkcialaj Ekvacioj. **50** (2007), 371–403.

24. Farwig, R., Neustupa, J.: On the spectrum of a Stokes-type operator arising from flow around a rotating body. Manuscripta Math. **122** (2007), 419–437.

25. Farwig, R., Nečasová, Š., Neustupa, J.: Spectral analysis of a Stokes-type operator arising from flow around a rotating body, J. Math. Soc. Japan, **63** (2011), 163–194.

26. Galdi, G. P.: On the asymptotic structure of D-solutions to the steady-state Navier-Stokes equations in exterior domains. In: Galdi, G. P. (ed.): Mathematical problems related to the Navier-Stokes equations. Advances in Mathematics in the Applied Sciences **11**. World Scientific, Singapore, p. 81–105.

27. Galdi, G.P.: An Introduction to the mathematical theory of the Navier-Stokes equations. Vol. I. Linearized steady problems (rev. ed.). Springer, New York e.a., 1998.

28. Galdi, G.P.: An introduction to the mathematical theory of the Navier-Stokes equations. Vol. II. Nonlinear steady problems. Springer, New York e.a., 1994.

29. Galdi, G. P.: "An introduction to the mathematical theory of the Navier-Stokes equations. Steady-state problems" (2nd ed.), Springer, New York e.a., 2011.

30. Galdi, G. P., Kyed, M.: Steady-state Navier-Stokes flows past a rotating body: Leray solutions are physically reasonable. Arch. Rat. Mech. Anal. **200** (2011), 21–58.

31. Galdi, G. P.: On the motion of a rigid body in a viscous liquid: A mathematical analysis with applications, Handbook of Mathematical Fluid Dynamics, Volume 1, Ed. by S. Friedlander, D. Serre, Elsevier (2002).

32. Galdi, G. P.: Steady flow of a Navier-Stokes fluid around a rotating obstacle. Essays and papers dedicated to the memory of Clifford Ambrose Truesdell III, Vol.II, J. Elasticity **71**, 1-3, (2003), 1–31.

33. Galdi, G. P., Silvestre, A. S.: Strong solutions to the Navier-Stokes equations around a rotating obstacle. Arch. Rat. Mech. Appl. **176** (2005), 331–350.

34. Galdi, G. P., Silvestre, S. A.: The steady motion of a Navier-Stokes liquid around a rigid body. Arch. Rat. Mech. Appl. **184** (2007), 371–400.

35. Geissert, M., Heck, H., Hieber, M.: L^p theory of the Navier-Stokes flow in the exterior of a moving or rotating obstacle. J. Reine A. Math. **596**, (2006), 45–62.

36. Guenther, R. B., Thomann, E. A.: The fundamental solution of the linearized Navier-Stokes equations for spinning bodies in three spatial dimensions – time dependent case. J. Math. Fluid Mech. **7** (2005), 1–22.

37. Hishida, T.: An existence theorem for the Navier-Stokes flow in the exterior of a rotating obstacle. Arch. Rational Mech. Anal. **150**, (1999), 307–348.

38. Hishida, T.: The Stokes operator with rotating effect in exterior domains. Analysis **19**, (1999), 51–67.

39. Hishida, T.: L^q estimates of weak solutions to the stationary Stokes equations around a rotating body. J. Math. Soc. Japan. **58** (2006), 743–767.

40. Hishida, T., Shibata, Y.: $L_p - L_q$ estimate of the Stokes operator and Navier-Stokes flows in the exterior of a rotating obstacle. RIMS Kokyuroku Bessatsu, B1, (2007), 167–188.

41. Kračmar, S., Novotný, A., Pokorný, M.: Estimates of Oseen kernels in weighted L^p spaces. J. Math. Soc. Japan **53** (2001), 59–111.

42. Kračmar, S., Penel, P.,: Variational properties of a generic model equation in exterior 3D domains. Funkcial. Ekv. **47** (2004), 499–523.

43. Kračmar, S., Penel, P., New regularity results for a generic model equation in exterior 3D domains. Banach Center Publications Warsaw **70** (2005), 139–155.

44. Kračmar, S., Nečasová, Š., Penel, P.: Estimates of weak solutions in anisotropically weighted Sobolev spaces to the stationary rotating Oseen equations. IASME Transactions **2**, (2005), 854–861.

45. Kračmar, S., Nečasová, Š., Penel, P.: Anisotropic L^2 estimates of weak solutions to the stationary Oseen type equations in R^3 for a rotating body. RIMS Kokyuroku Bessatsu. B1, (2007), 219–235.

46. Kračmar, S., Nečasová, Š., Penel, P.: Anisotropic L^2 estimates of weak solutions to the stationary Oseen type equations in 3D - exterior domain for a rotating body. J. Math. Soc. Japan **62** (2010), 1, 239–268.

47. Kyed, M.: On the asymptotic structure of a Navier-Stokes flow past a rotating body. J. Math. Soc. Japan, **66** (2014), 1–16.

48. Kyed, M.: Periodic solutions to the Navier-Stokes equations. habilitation thesis, Technische Universität Darmstadt, Darmstadt, 2012.

49. Da Prato, G., Lunardi, A.: On the Ornstein-Uhlenbeck operator in spaces of continuous functions, J. Funct. Anal., **131** (1995), 94–114.

50. Magnus, W., Oberhettinger, F., Soni, R. P.: Formulas and theorems for the special functions of mathematical physics (3rd ed.). Die Grundlehren der mathematischen Wissenschaften, Band 52. Springer, New York e. a., 1966.

51. Nečas, J.: Les méthodes directes en théorie des équations elliptiques, Masson, Paris, 1967.

52. Nečasová, Š., Schumacher, K.: Strong solution to the Stokes equations of a flow around a rotating body in weighted L^q spaces. Math. Nachr. **284** (2011), 13, 1701–1714.

53. Nečasová, Š.: On the problem of the Stokes flow and Oseen flow in $\mathbb{R}^3$ with Coriolis force arising from fluid dynamics. IASME Transaction **2** (2005), 1262–1270.

54. Nečasová, Š. : Asymptotic properties of the steady fall of a body in viscous fluids. Math. Meth. Appl. Sci. **27**, (2004), 1969–1995.

55. Sazonov, L. I.: On the asymptotics of the solution to the three-dimensional problem of flow far from streamlined bodies (Translation). Izvestiya: Math. **59** (1995), 1051–1075.

56. Solonnikov, V. A.: A priori estimates for second order parabolic equations. Trudy Mat. Inst. Steklov. **70** (1964), 133-212 (Russian); English translation: AMS Translations **65** (1967), 51–137.

57. Weinberger, H. F.: Variational properties of steady fall in Stokes flow. J. Fluid Mech.. **52**, 2, (1972), 321–344.

58. Weinberger, H. F.: On the steady fall of a body in a Navier-Stokes fluid. Partial differential equations. (Proc. Sympos. Pure Math., Vol. XXIII, Univ. California, Berkeley, Calif., (1971), 421–439. Amer. Math. Soc., Providence, R. I., 1973.

Index

© Atlantis Press and the author(s) 2016
Š. Nečasová and S. Kračmar, *Navier–Stokes Flow Around a Rotating Obstacle*,
Atlantis Briefs in Differential Equations 3, DOI 10.2991/978-94-6239-231-1

Printed in the United States
By Bookmasters